Ground Coastal Environments
Problems and Perspectives

Ground Coastal Environments

Problems and Perspectives

Chinmay Chatterjee

RANDOM PUBLICATIONS

NEW DELHI - 110 002 (INDIA)

Ground Coastal Environments: Problems and Perspectives

ISBN 978-93-51112-70-9

Published in 2014 in India by

RANDOM PUBLICATIONS

4376-A/4B, Gali Murari Lal, Ansari Road
New Delhi-110 002
Phone: +9111-43580356, 23289044
E-mail: randomexports@gmail.com; sales@randompublications.com; info@randompublications.com

Reprint 2021

Type Setting by: Friends Media, Delhi-110089
Printed at : Replika Press Pvt. Ltd.

Preface

Planning, siting, design, and construction of coastal residential buildings require an understanding of the coastal environment—including a basic understanding of coastal geology, coastal processes, regional variations in coastline characteristics, and coastal sediment budgets. Coastal processes interact with the local coastal geology and sediment supply to form and modify the physical features that will be referred to frequently in this course: beaches, dunes, bluffs, and upland areas. Water levels, waves, currents, and winds will vary with time at a given location (sometimes according to short-term, seasonal, or longer-term patterns) and will vary geographically at any point in time. A good analogy is weather: weather conditions at a given location undergo significant variability over time but tend to follow seasonal and other patterns. Further, weather conditions can differ substantially from one location to another at the same point in time. The main causes of this degradation, apart from natural disasters, are poverty and the pressures of economic development at local to global scales. Economic gains, many bringing only short-term benefits, are being made at the expense of the integrity of ecosystems and the vulnerable communities that they support. The overexploitation of offshore fisheries impacts on the food security of coastal populations. Another key concern is the modification of river flows to the coast by damming and irrigation, and pollution from land, marine and atmospheric sources.

The coastal region occupies some of the most potential aquifer systems of the country. The coastal aquifers of India ranges from that of Jurassic to Recent and is seen almost all along the coast right from Gujarat to West Bengal. Some of the aquifers especially the Tertiary to Recent ones are highly potential and are developed extensively. The small island aquifers of Lakshadweep are highly sensitive since fresh water is seen floating as a thin lens over sea water here. The problems are also complex in the coastal area. Some of them are sea water intrusion, salinity from the aquifer (in situ) material, pollution, global warming and its impact on these aquifer systems. The hydogeologic scenario of the coastal area is discussed briefly in the paper. Sea water intrusion is reported in Gujarat and Tamil Nadu and the salinity in other areas are mostly derived from the aquifer materials. The sea level changes in the past and the present day sea level rise due to global warming and the impact on the coastal aquifers especially that of the small island aquifers of Lakshadweep are discussed in detail.

The present book has been designed to outline the basic and fundamental aspects of this subject to be understood in its right perspective. The book uses straight forward, less-technical jargon and manages to introduce each chapter with a basic concept, which ultimately evolves into a more specific detailed principle.

I thank all members of my team who have helped in the preparation of the book. My special thanks go to "Random Publications" who have published the book.

—Chinmay Chatterjee

Contents

1

Introduction

Coast

A coastline or seashore is the area where land meets the sea or ocean. A precise line that can be called a coastline cannot be determined due to the dynamic nature of tides. The term "coastal zone" can be used instead, which is a spatial zone where interaction of the sea and land processes occurs. Both the terms coast and coastal are often used to describe a geographic location or region; for example, New Zealand's West Coast, or the East and West Coasts of the United States.

A pelagic coast refers to a coast which fronts the open ocean, as opposed to a more sheltered coast in a gulf or bay. A shore, on the other hand, can refer to parts of the land which adjoin any large body of water, including oceans (sea shore) and lakes (lake shore). Similarly, the somewhat related term "bank" refers to the land alongside or sloping down to a river (riverbank) or to a body of water smaller than a lake. "Bank" is also used in some parts of the world to refer to an artificial ridge of earth intended to retain the water of a river or pond; in other places this may be called a levee.

While many scientific experts might agree on a common definition of the term "coast", the delineation of the extents of a coast differ according to jurisdiction, with many scientific and government authorities in various countries differing for economic and social policy reasons.

Formation

Tides often determine the range over which sediment is deposited or eroded. Areas with high tidal ranges allow waves to reach farther up the shore, and areas with lower tidal ranges produce deprossosition at a smaller elevation interval. The tidal range is influenced by the size and shape of the coastline. Tides do not typically cause erosion by themselves; however, tidal bores can erode as the waves surge up river estuaries from the ocean.

Waves erode coastline as they break on shore releasing their energy; the larger the wave the more energy it releases and the more sediment it moves. Coastlines with longer shores have more room for the waves to disperse their energy, while coasts with cliffs and short shore faces give little room for the wave energy to be dispersed. In these areas the wave energy breaking against the cliffs is higher, and air and water are compressed into cracks in the rock, forcing the rock apart, breaking it down. Sediment deposited by waves comes from eroded cliff faces and is moved along the coastline by the waves. This forms an abrasion or cliffed coast.

Sediment deposited by rivers is the dominant influence on the amount of sediment located on a coastline. Today riverine deposition at the coast is often blocked by dams and other human regulatory devices, which remove the sediment from the stream by causing it to be deposited inland.

Like the ocean which shapes them, coasts are a dynamic environment with constant change. The Earth's natural processes, particularly sea level rises, waves and various weather phenomena, have resulted in the erosion, accretion and reshaping of coasts as well as flooding and creation of continental shelves and drowned river valleys (rias).

Environmental Importance

The coast and its adjacent areas on and off shore are an important part of a local ecosystem: the mixture of fresh water and salt water in estuaries provides many nutrients for marine life. Salt marshes and beaches also support a diversity of plants, animals and insects crucial to the food chain.

The high level of biodiversity creates a high level of biological activity, which has attracted human activity for thousands of years.

Human Impacts

Human Uses of Coasts: More and more of the world's people live in coastal regions. Many major cities are on or near good harbours and have port facilities. Some landlocked places have achieved port status by building canals.

The coast is a frontier that nations have typically defended against military invaders, smugglers and illegal migrants. Fixed coastal defences have long been erected in many nations and coastal countries typically have a navy and some form of coast guard.

Coasts, especially those with beaches and warm water, attract tourists. In many island nations such as those of the Mediterranean, South Pacific and Caribbean, tourism is central to the economy. Coasts offer recreational activities such as swimming, fishing, surfing, boating, and sunbathing. Growth management can be a challenge for coastal local authorities who often struggle to provide the infrastructure required by new residents.

Threats to a Coast

Coasts also face many human-induced environmental impacts. The human influence on climate change is thought to contribute to an accelerated trend in sea level rise which threatens coastal habitats.

Pollution can occur from a number of sources: garbage and industrial debris; the transportation of petroleum in tankers, increasing the probability of large oil spills; small oil spills created by large and small vessels, which flush bilge water into the ocean.

Fishing has declined due to habitat degradation, overfishing, trawling, bycatch and climate change. Since the growth of global fishing enterprises after the 1950s, intensive fishing has spread from a few concentrated areas to encompass nearly all fisheries. The scraping of the ocean floor in bottom dragging is devastating to coral, sponges and other long-lived species that do not recover quickly. This destruction alters the functioning of the ecosystem and can permanently alter species composition and biodiversity. Bycatch, the capture of unintended species in the course of fishing, is typically returned to the ocean only to die from injuries or exposure. Bycatch represents about a quarter of all marine catch. In the case of shrimp capture, the bycatch is five times larger than the shrimp caught.

It is believed that melting Arctic ice will cause sea levels to rise and flood costal areas.

Conservation

Extraordinary population growth in the 20th century has placed stress on the planet's ecosystems. For example, on Saint Lucia, harvesting mangrove for timber and clearing for fishing reduced the mangrove forests, resulting in a loss of habitat and spawning grounds for marine life that was unique to the area. These forests also helped to stabilize the coastline. Conservation efforts since the 1980s have partially restored the ecosystem.

Types of Coast

According to one principle of classification, an emergent coastline is a coastline which has experienced a fall in sea level, because of either a global sea level change, or local uplift. Emergent coastlines are identifiable by the coastal landforms, which are above the high tide mark, such as raised beaches. In contrast, a submergent coastline is one where the sea level has risen, due to a global sea level change, local subsidence, or isostatic rebound. Submergent coastlines are identifiable by their submerged, or "drowned" landforms, such as rias (drowned valleys) and fjords.

According to a second principle of classification, a concordant coastline is a coastline where bands of different rock types run parallel to the shore. These rock types are usually of varying resistance, so the coastline forms distinctive landforms, such as coves. Discordant coastlines feature distinctive landforms because the rocks are eroded by ocean waves. The less resistant rocks erode faster, creating inlets or bays; the more resistant rocks erode more slowly, remaining as headlands or outcroppings.

Other coastal categories:

- A cliffed coast or abrasion coast is one where marine action has produced steep declivities known as cliffs.
- A flat coast is one where the land gradually descends into the sea.
- A graded shoreline is one where wind and water action has produced a flat and straight coastline.

Emergent Coastline

Emergent coastlines are stretches along the coast that have been exposed by the sea due to a relative fall in sea levels. This occurs due to either isostasy or eustasy.

Emergent coastline are the opposite of submergent coastlines which have experienced a relative rise in sea-levels.

The specific landform of an emergent coastline may be:

- Raised beach or machair
- Wave cut platform
- Sea cave such as King's Cave, Isle of Arran

The Scottish Gaelic word machair or machar refers to a fertile low-lying raised beach found on the some of the coastlines of Ireland and Scotland, in particular the Outer Hebrides. Hudson Bay, in Canada's north, is an example of an emergent coastline. Currently it is still emerging by as much as one cm/year.

Submergent Coastline

Submergent coastlines are stretches along the coast that have been inundated by the sea due to a relative rise in sea levels. This occurs due to either isostacy or eustacy.

Submergent coastline are the opposite of emergent coastlines which have experienced a relative fall in sea-levels.

Features of a submergent coastline are:

- drowned river valleys or rias
- drowned glaciated valleys or fjords.

Estuaries are often the drowned mouths of rivers.

Concordant Coastline

A concordant, longitudinal, or Pacific type coastline occurs where beds, or layers, of differing rock types are folded into ridges that run parallel to the coast. The outer hard rock (for example, granite) provides a protective barrier to erosion of the softer rocks (for example, clays) further inland. Sometimes the outer hard rock is punctured, allowing the sea to erode the softer rocks behind. This creates a cove, a circular area of water with a relatively narrow entrance from the sea. Lulworth Cove in Dorset is situated on a concordant coastline.

The outer hard rock is Portland limestone. The sea has broken through this barrier and easily eroded the clays behind it. A chalk cliff face at the back of the cove slows further erosion. Erosion is just starting to the west, where the sea has again broken through the Portland limestone barrier at Stair Hole.

The concordant coast may take one of two landform types. The Dalmatian type, named from Dalmatia on the Adriatic Sea, features long offshore islands and coastal inlets that are parallel to the coastline. The Adriatic Sea itself is a concordant landform, consisting of a body of water between parallel ranges. The second landform is the Haff type as in the Haffs, or lagoons, of the southern shore of the Baltic Sea, which are enclosed by long spits of sand parallel to the low coast.

The converse of concordant coastline is a discordant coastline.

Discordant Coastline

A discordant coastline occurs where bands of differing rock type run perpendicular to the coast. The differing resistance to erosion leads to the formation of headlands and bays. A hard rock type such as granite is resistant to erosion and creates a promontory whilst a softer rock type such as the clays of Bagshot Beds is easily eroded creating a bay.

Part of the Dorset coastline running north from the Portland limestone of Durlston Head is a clear example of a discordant coastline. The Portland limestone is resistant to erosion; then to the north there is a bay at Swanage where the rock type is a softer greensand. North of Swanage, the chalk outcrop creates the headland which includes Old Harry Rocks.

The converse of a discordant coastline is a concordant coastline.

Headlands and Bays

Headlands and bays are two related features of the coastal environment.

Geology and Geography

Headlands and bays are often found on the same coastline. A bay is surrounded by land on three sides, whereas a headland is surrounded by water on three sides. Headlands are characterized by high, breaking waves, rocky shores, intense erosion, and steep sea cliffs. Bays generally have less wave (and often wind) activity than the water outside the bay, and typically have sandy beaches. Headlands and bays form on discordant coastlines, where bands of rock of alternating resistance run perpendicular to the coast. Bays form where weak (less resistant) rocks (such as sands and clays) are eroded, leaving bands of stronger (more resistant) rocks (such as chalk, limestone, granite) forming a headland, or peninsula. This difference in the rate of erosion is caused by differential erosion.

Refraction of waves occurs on headlands concentrating wave energy on them, so many other landforms, such as caves, natural arches and stacks, form on headlands. Wave energy is directed at right angles to the wave crest and lines drawn at right angles to the wave crest (orthogonals) represent the direction of energy expenditure. Orthogonals converge on headlands and diverge in bays which concentrates wave energy on the headlands and dissipating wave energy in the bays. In the formation of sea cliffs, wave erosion undercuts the slopes at the shoreline and they retreat landward. This increases the shear stress in the cliff-forming material and accelerates mass movement. The debris from these landslides collects at the base of the cliff and are also removed by the waves, usually during storms where wave energy is greatest. This debris provides sediment, transported through longshore current for the nearby bay. Joints in the headlands are eroded back to form caves which erode further to form arches.

These gaps eventually collapse and leave tall stacks at the ends of the headlands. Eventually these too are eroded by the waves. Wave refraction disperses wave energy through the bay, and along with the sheltering effect of the headlands this protects bays from storms. This effect means that the waves reaching the shore in a bay are weaker than the waves reaching the headland and the bay is thus a safer place for water activities like surfing or swimming. Through the deposition of sediment within the bay and the erosion of the headlands, coastlines eventually straighten out then start the same process all over again.

Beach Stability

Beaches are dynamic geologic features that can fluctuate between advancement and retreat of sediment. The natural agents of fluctuation include waves, tides, currents, and winds. Man-made elements such as the interruption of sediment supply, such as a dam, and withdrawal of fluid can also affect beach stabilization. A headland bay beach can be classified as being in three different states of sedimentation. *Static equilibrium* refers to a beach that is stable and does not experience littoral drift or sediment deposition or erosion. Waves generally diffract around the headland(s) and near the beach when the beach is in a state of static equilibrium. *Dynamic equilibrium* occurs when the beach sediments are deposited and eroded at approximately equal rates. Beaches that have dynamic equilibrium are usually near a river that supplies sediment and would otherwise erode away without the river supply. *Unstable* beaches are usually a result of human interaction, such as a breakwater or dammed river. Unstable beaches are reshaped by continual erosion or deposition and will continue to erode or deposit until a state of equilibrium is reached in the bay.

Cove

A cove is a small type of bay or coastal inlet. Coves usually have narrow, restricted entrances, are often circular or oval, and are often situated within a larger bay. Small, narrow, sheltered bays, inlets, creeks, or recesses in a coast are often considered coves. Colloquially, the term can be used to describe a sheltered bay.

Geomorphology describes coves as precipitously walled and rounded cirque-like openings as in a valley extending into or down a mountainside, or in a hollow or nook of a cliff or steep mountainside.

Coves are formed by differential erosion. Differential erosion occurs when softer rocks are worn away faster than the harder rocks surrounding them. These rocks further erode to form a circular bay with a narrow entrance called a cove.

Figure: *McWay Cove, California, USA*

An example of a cove is Lulworth Cove on the Jurassic Coast in Dorset, England. West of it a second cove, Stair Hole, is forming.

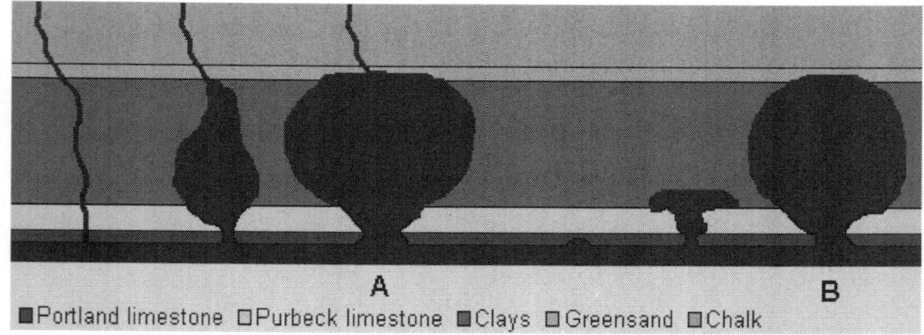

Figure: *Two examples of how coves form. The rock types are those of Lulworth Cove. In example A a river breaks through the resistant chalk back rock and limestone leaving the weak clays to be rapidly eroded. In example B the sea breaks through the limestone, perhaps by forming a cave, and then erodes the clay away.*

Peninsula

A peninsula is a piece of land that is bordered by water on three sides but connected to mainland. The surrounding water is usually understood to belong to a single, contiguous body, but is not always explicitly defined as such. In many Germanic and Celtic languages and also in Baltic, Slavic, Hungarian, Chinese, Hebrew and Korean, peninsulas are called "half-islands". A peninsula can also be a headland (head), cape, island promontory, bill, point, or spit. Note that a point is generally considered a tapering piece of land projecting into a body of water that is less prominent than a cape. In English, the plural of peninsula is *peninsulas* or, less commonly, *peninsulae.*

Estuary

An estuary is a partly enclosed coastal body of brackish water with one or more rivers or streams flowing into it, and with a free connection to the open sea. Estuaries form a transition zone between river environments and maritime environments and are subject to both marine influences, such as tides, waves, and the influx of saline water; and riverine influences, such as flows of fresh water and sediment. The inflows of both sea water and fresh water provide high levels of nutrients in both the water column and sediment, making estuaries among the most productive natural habitats in the world.

Most existing estuaries were formed during the Holocene epoch by the flooding of river-eroded or glacially scoured valleys when the sea level began to rise about 10,000-12,000 years ago. Estuaries are typically classified by their geomorphological features or by water circulation patterns and can be referred to by many different names, such as bays, harbours, lagoons, inlets, or sounds, although some of these water bodies do not strictly meet the above definition of an estuary and may be fully saline.

The banks of many estuaries are amongst the most heavily populated areas of the world, with about 60% of the world's population living along estuaries and the coast. As a result, many estuaries are suffering degradation by many factors, including sedimentation from soil erosion from deforestation, overgrazing, and other poor farming practices; overfishing; drainage and filling of wetlands; eutrophication due to excessive nutrients from sewage and animal wastes; pollutants including heavy metals, polychlorinated biphenyls, radionuclides and hydrocarbons from sewage inputs; and diking or damming for flood control or water diversion.

Definition

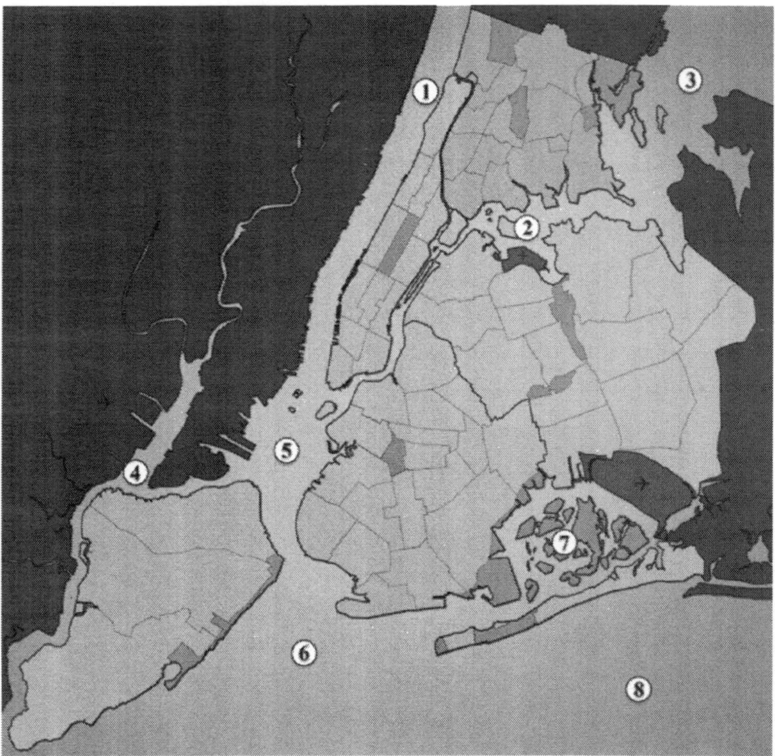

Figure: *Hudson River estuary waterways around New York City, USA: 1. Hudson River, 2. East River, 3. Long Island Sound, 4. Newark Bay, 5. Upper New York Bay, 6. Lower New York Bay, separated from Upper New York Bay by the Narrows strait, 7. Jamaica Bay, and 8. Atlantic Ocean.*

The word "estuary" is derived from the Latin word *aestuarium* meaning tidal inlet of the sea, which in itself is derived from the term *aestus*, meaning tide. There have been many definitions proposed to describe an estuary. The most widely accepted definition is: "a semi-enclosed coastal body of water, which has a free connection with the open sea, and within sea water is measurably diluted with freshwater derived from land drainage". However, this definition excludes a number of coastal water bodies such as coastal lagoons and brackish seas. A more comprehensive definition of an estuary is "a semi-enclosed body of water connected to the sea as far as the tidal limit or the salt intrusion limit and receiving freshwater runoff; however the freshwater inflow may not be perennial, the connection to the sea may be closed for part of the year and tidal influence may be negligible". This definition includes classical estuaries as well as fjords, lagoons, river mouths, and tidal creeks. An estuary is a dynamic ecosystem with a connection with the open sea through which the sea water enters with the rhythm of the tides. The sea water entering the estuary is diluted by the fresh water flowing from rivers and streams. The pattern of dilution varies between different estuaries and depends on the volume of fresh water, the tidal range, and the extent of evaporation of the water in the estuary.

Classification Based on Geomorphology

Drowned river valleys: Their width-to-depth ratio is typically large, appearing wedge-shaped in the inner part and broadening and deepening seaward. Water depths rarely exceed 30 m (100 ft). Examples of this type of estuary in the U.S. are the Hudson River, Chesapeake Bay, and Delaware Bay along the Mid-Atlantic coast; and along the Gulf coast, Galveston Bay and Tampa Bay.

Lagoon-type or Bar-built

These estuaries are semi-isolated from ocean waters by barrier beaches (barrier islands and barrier spits). Formation of barrier beaches partially encloses the estuary, with only narrow inlets allowing contact with the ocean waters. Bar-built estuaries typically develop on gently sloping plains located along tectonically stable edges of continents and marginal sea coasts. They are extensive along the Atlantic and Gulf coasts of the U.S. in areas with active coastal deposition of sediments and where tidal ranges are less than 4 m (13 ft). The barrier beaches that enclose bar-built estuaries have been developed in several ways:

- building up of offshore bars by wave action, in which sand from the sea floor is deposited in elongated bars parallel to the shoreline,
- reworking of sediment discharge from rivers by wave, current, and wind action into beaches, overwash flats, and dunes,
- engulfment of mainland beach ridges (ridges developed from the erosion of coastal plain sediments around 5000 years ago) due to sea level rise and resulting in the breaching of the ridges and flooding of the coastal lowlands, forming shallow lagoons, and
- elongation of barrier spits from the erosion of headlands due to the action of longshore currents, with the spits growing in the direction of the littoral drift.

Barrier beaches form in shallow water and are generally parallel to the shoreline, resulting in long, narrow estuaries. The average water depth is usually less than 5 m (16 ft), and rarely exceeds 10 m (33 ft). Examples of bar-built estuaries are Barnegat Bay, New Jersey; Laguna Madre, Texas; and Pamlico Sound, North Carolina.

Fjord-type

Fjord-type estuaries are formed in deeply eroded valleys formed by glaciers. These U-shaped estuaries typically have steep sides, rock bottoms, and underwater sills contoured by glacial movement. The estuary is shallowest at its mouth, where terminal glacial moraines or rock bars form sills that restrict water flow. In the upper reaches of the estuary, the depth can exceed 300 m (1,000 ft). The width-to-depth ratio is generally small. In estuaries with very shallow sills, tidal oscillations only affect the water down to the depth of the sill, and the waters deeper than that may remain stagnant for a very long time, so there is only an occasional exchange of the deep water of the estuary with the ocean. If the sill depth is deep, water circulation is less restricted, and there is a slow but steady exchange of water between the estuary and the ocean. Fjord-type estuaries can be found along the coasts of Alaska, the Puget Sound region of western Washington state, British Columbia, eastern Canada, Greenland, Iceland, New Zealand, and Norway.

Tectonically Produced

These estuaries are formed by subsidence or land cut off from the ocean by land movement associated with faulting, volcanoes, and landslides. Inundation from eustatic sea level rise during the Holocene Epoch has also contributed to the formation of these estuaries. There are only a small number of tectonically produced estuaries; one example is the San Francisco Bay, which was formed by the crustal movements of the San Andreas fault system causing the inundation of the lower reaches of the Sacramento and San Joaquin rivers.

Classification Based on Water Circulation

Salt Wedge: In this type of estuary, river output greatly exceeds marine input and tidal effects have a minor importance. Fresh water floats on top of the seawater in a layer that gradually thins as it moves seaward. The denser seawater moves landward along the bottom of the estuary, forming a wedge-shaped layer that is thinner as it approaches land. As a velocity difference develops between the two layers, shear forces generate internal waves at the interface, mixing the seawater upward with the freshwater. An example of a salt wedge estuary is the Mississippi River.

Partially Mixed

As tidal forcing increases, river output becomes less than the marine input. Here, current induced turbulence causes mixing of the whole water column such that salinity varies more longitudinally rather than vertically, leading to a moderately stratified condition. Examples include the Chesapeake Bay and Narragansett Bay.

Vertically Homogenous

Tidal mixing forces exceed river output, resulting in a well mixed water column and the disappearance of the vertical salinity gradient. The freshwater-seawater boundary is eliminated due to the intense turbulent mixing and eddy effects. The lower reaches of the Delaware Bay and the Raritan River in New Jersey are examples of vertically homogenous estuaries.

Inverse

Inverse estuaries occur in dry climates where evaporation greatly exceeds the inflow of fresh water. A salinity maximum zone is formed, and both riverine and oceanic water flow close to the surface towards this zone. This water is pushed downward and spreads along the bottom in both the seaward and landward direction. An example of an inverse estuary is Spencer Gulf, South Australia.

Intermittent

Estuary type varies dramatically depending on freshwater input, and is capable of changing from a wholly marine embayment to any of the other estuary types.

Physiochemical Variation

The most important variable characteristics of estuary water are the concentration of dissolved oxygen, salinity and sediment load. There is extreme spatial variability in salinity, with a range of near zero at the tidal limit of the tributary river(s) to 3.4% at the estuary mouth. At any one point the salinity will vary considerably over time and seasons, making it a harsh environment for organisms. Sediment often settles in intertidal mudflats which are extremely difficult to colonize. No points of attachment exist for algae, so vegetation based habitat is not established. Sediment can also clog feeding and respiratory structures of species, and special adaptations exist within mudflat species to cope with this problem. Lastly, dissolved oxygen variation can cause problems for life forms. Nutrient-rich sediment from man-made sources can promote primary production life cycles, perhaps leading to eventual decay removing the dissolved oxygen from the water; thus hypoxic or anoxic zones can develop.

Implications for Marine Life

Estuaries provide habitats for a large number of organisms and support very high productivity. Estuaries provide habitats for many fish nurseries, depending upon their locations in the world, such as salmon and sea trout. Also, migratory bird populations, such as the black-tailed godwit, *Limosa limosa islandica* make essential use of estuaries.

Two of the main challenges of estuarine life are the variability in salinity and sedimentation. Many species of fish and invertebrates have various methods to control or conform to the shifts in salt concentrations and are termed osmoconformers and osmoregulators. Many animals also burrow to avoid predation and to live in the more stable sedimental environment. However, large numbers of bacteria are found within the sediment which have a very high oxygen demand. This reduces the levels of oxygen within the sediment often resulting in partially anoxic conditions, which can be further exacerbated by limited water flux.

Phytoplankton are key primary producers in estuaries. They move with the water bodies and can be flushed in and out with the tides. Their productivity is largely dependant upon the turbidity of the water. The main phytoplankton present are diatoms and dinoflagellates which are abundant in the sediment. It is important to remember that a primary source of food for many organisms on estuaries, including bacteria, is detritus from the settlement of the sedimentation.

Human Impact

Of the thirty-two largest cities in the world, twenty-two are located on estuaries. For example, New York City is located at the mouth of the Hudson River estuary.

As ecosystems, estuaries are under threat from human activities such as pollution and overfishing. They are also threatened by sewage, coastal settlement, land clearance and much more. Estuaries are affected by events far upstream, and concentrate materials such as pollutants and sediments. Land run-off and industrial, agricultural, and domestic waste enter rivers and are discharged into estuaries. Contaminants can be introduced which do not disintegrate rapidly in the marine environment, such as plastics, pesticides, furans, dioxins, phenols and heavy metals.

Such toxins can accumulate in the tissues of many species of aquatic life in a process called bioaccumulation. They also accumulate in benthic environments, such as estuaries and bay muds: a geological record of human activities of the last century.

For example, Chinese and Russian industrial pollution, such as phenols and heavy metals, has devastated fish stocks in the Amur River and damaged its estuary soil. Estuaries tend to be naturally eutrophic because land runoff discharges nutrients into estuaries. With human activities, land run-off also now includes the many chemicals used as fertilizers in agriculture as well as waste from livestock and humans. Excess oxygen depleting chemicals in the water can lead to hypoxia and the creation of dead zones. This can result in reductions in water quality, fish, and other animal populations.

Overfishing also occurs. Chesapeake Bay once had a flourishing oyster population which has been almost wiped out by overfishing. Historically the oysters filtered the estuary's entire water volume of excess nutrients every three or four days. Today that process takes almost a year, and sediment, nutrients, and algae can cause problems in local waters. Oysters filter these pollutants, and either eat them or shape them into small packets that are deposited on the bottom where they are harmless.

River Delta

A river delta is a landform that is formed at the mouth of a river, where the river flows into an ocean, sea, estuary, lake, or reservoir. Deltas are formed from the deposition of the sediment carried by the river as the flow leaves the mouth of the river. Over long periods of time, this deposition builds the characteristic geographic pattern of a river delta.

Despite a popular legend to the contrary, this usage of the word *delta* was not coined by Herodotus.

Formation

River deltas form when a river carrying sediment reaches either

(1) a body of standing water, such as a lake, ocean, or reservoir,

(2) another river that cannot remove the sediment quickly enough to stop delta formation, or

(3) an inland region where the water spreads out and deposits sediments.

The tidal currents also cannot be too strong as the sediments would be washed out into the water body faster than the river can deposit the sediments. And of course the river must be carrying enough sediments to be layered into deltas over time. The rivers velocity decreases rapidly causing it to deposit the majority, if not all, of its load. This aluvium builds up to form the river delta. When the flow enters the standing water, it is no longer confined to its channel and expands in width. This flow expansion results in a decrease in the flow velocity, which diminishes the ability of the flow to transport sediment. As a result, sediment drops out of the flow and deposits. Over time, this single channel will build a deltaic lobe (such as the bird's-foot of the Mississippi or Ural River deltas), pushing its mouth further into the standing water.

As the deltaic lobe advances, the gradient of the river channel becomes lower because the river channel is longer but has the same change in elevation. As the slope of the river channel decreases, it becomes unstable for two reasons. First, the water under the force of gravity will tend to flow in the most direct course down slope. If the river breaches its natural levees (i.e., during a flood), it will spill out onto a new course with a shorter route to the ocean, thereby obtaining a more stable steeper slope. Second, as its slope gets lower, the amount of shear stress on the bed will decrease, which will result in deposition of sediment within the channel and a rise in the channel bed relative to the floodplain.

This will make it easier for the river to breach its levees and cut a new channel that enters the body of standing water at a steeper slope. Often when the channel does this, some of its flow can remain in the abandoned channel. When these channel switching events occur a mature delta will gain a distributary network.

Another way in which these distributary networks may form is from the deposition of mouth bars (mid-channel sand and/or gravel bars at the mouth of a river). When this mid-channel bar is deposited at the mouth of a river, the flow is routed around it. This results in additional deposition on the upstream end of the mouth-bar, which splits the river into two distributary channels. A good example of the result of this process is the Wax Lake Delta in Louisiana.

In both of these cases, depositional processes force redistribution of deposition from areas of high deposition to areas of low deposition. This results in the smoothing of the planform (or map-view) shape of the delta as the channels move across its surface and deposit sediment. Because the sediment is laid down in this fashion, the shape of these deltas approximates a fan. It is closer to an ideal fan the more often the flow changes course because more rapid changes in channel position results in more uniform deposition of sediment on the delta front.

The Mississippi and Ural River deltas, with their bird's-feet, are examples of rivers that do not avulse often enough to form a symmetrical fan shape. Alluvial fan deltas, avulse frequently and more closely approximate an ideal fan shape.

Types of Deltas

Deltas are typically classified according to the main control on deposition, which is usually either a river, waves, or tides. These controls have a large effect on the shape of the resulting delta.

Wave-dominated Deltas

In wave dominated deltas, wave erosion controls the shape of the delta, and much of the sediment emanating from the river mouth is deflected along the coast line. Deltas of this form, such as the Nile Delta, tend to have a characteristic Greek-capital-delta shape (Δ).

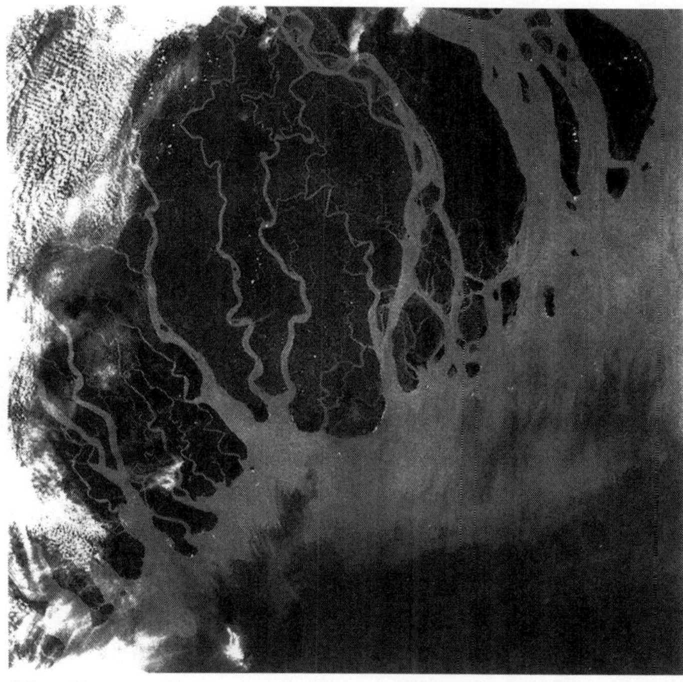

Figure: *The Ganges Delta in India and Bangladesh is the largest delta in the world and it is also one of the most fertile regions in the world.*

Tide-dominated Deltas

Erosion is also an important control in tide dominated deltas, such as the Ganges Delta, which may be mainly submarine, with prominent sand bars and ridges. This tends to produce a "dendritic" structure. Tidal deltas behave differently from river- and wave-dominated deltas, which tend to have a few main distributaries. Once a wave- or river- distributary silts up, it is abandoned, and a new channel forms elsewhere. In a tidal delta, new distributaries are formed during times when there's a lot of water around – such as floods or storm surges. These distributaries slowly silt up at a pretty constant rate until they fizzle out.

Gilbert Deltas

A Gilbert delta (named after Grove Karl Gilbert) is a specific type of delta that is formed by coarse sediments, as opposed to gently-sloping muddy deltas such as that of the Mississippi. For example, a mountain river depositing sediment into a freshwater lake would form this kind of delta. While some authors describe both lacustrine and marine locations of Gilbert deltas, others note that their formation is more characteristic of the freshwater lakes, where it is easier for the river water to mix with the lakewater faster (as opposed to the case of a river falling into the sea or a salt lake, where less dense fresh water brought by the river stays on top longer).

G.K. Gilbert himself first described this type of delta on Lake Bonneville in 1885. Elsewhere, similar structures can be found e.g. at the mouths of several creeks flowing into Okanagan Lake in British Columbia and forming prominent peninsulas at Naramata (49°352 303 N 119°352 303 W), Summerland (49°342 233 N 119°372 453 W), or Peachland (49°472 003 N 119°422 453 W)

Estuaries

Other rivers, particularly those located on coasts with significant tidal range, do not form a delta but enter into the sea in the form of an estuary. Notable examples include the Saint Lawrence River and the Tagus estuary.

Inland Deltas

In rare cases the river delta is located inside a large valley and is called an inverted river delta. Sometimes a river will divide into multiple branches in an inland area, only to rejoin and continue to the sea; such an area is known as an inland delta, and often occur on former lake beds. The Inner Niger Delta and Peace–Athabasca

Delta are notable examples. The Amazon has also an inland delta before the island of Marajó.

Figure: *Okavango Delta*

In some cases a river flowing into a flat arid area splits into channels which then evaporate as it progresses into the desert. Okavango Delta in Botswana is one well-known example.

Sedimentary Structure

The formation of a delta consists of three main forms: the topset, foreset/frontset, and bottomset.

- The bottomset beds are created from the suspended sediment that settles out of the water as the river flows into the body of water and loses energy. The suspended load is carried out the furthest into the body of water than all other types of sediment creating a turbidite. These beds are laid down in horizontal layers and consist of smaller grains.
- The foreset beds in turn build over the bottomset beds as the main delta form advances. The foreset beds consist of the bed load that the river is moving along which consists of larger sediments that roll along the main channel. When it reaches the edge of the form, the bed load rolls over the edge, and builds up in steeply angled layers over the top of the bottomset beds. The angle of the outermost edge of the delta is created

by the sediments angle of repose. As the foresets build outward (which make up the majority of the delta) they pile up and miniature landslides occur. This slope is created in this fashion as the bedload continues to be deposited and the delta moves outward. In cross section, one would see the foresets lying in angled, parallel bands, showing each stage of the creation of the delta.

- The topset beds in turn overlay the foresets, and are horizontal layers of smaller sediment size that form as the main channel of the river shifts elsewhere and the larger particles of the bed load no longer are deposited. As the channels move across the top of the delta, the suspended load settles out in horizontal beds over the top. The topset bed is subdivided into two regions: the upper delta plain and the lower delta plain. The upper delta plain is unaffected by the tide, while the boundary with the lower delta plain is defined by the upper limit of tidal influence.

Deltas and Alluvial Fans

Deltas are differentiated from alluvial fans in that deltas have a shallow slope, contain fine-grained sediment (sand and mud), and generally always flow into a body of water (with the exception of inland deltas). Alluvial fans, on the other hand, are steep, have coarse-grained sediments (including boulders), and are dominated by debris flows and large floods; these floods are often flash floods. They can either flow onto a land surface, or into a body of water; in the latter case, they are called alluvial fan deltas.

Examples of Deltas

The most famous delta is that of the Nile River, and it is this delta from which the term is derived. The Ganges/Brahmaputra combination delta spans most of Bangladesh and West Bengal, empties into the Bay of Bengal and is the world's largest delta. Other rivers with notable deltas include, the Fly River, the Kaveri, the Niger River, the Tigris-Euphrates, the Rhine, the Po, the Rhône, the Danube, the Ebro, the Volga, the Lena, the Indus, the Ayeyarwady (Irrawaddy), the Mekong, the Huanghe, the Yangtze, the Sacramento-San Joaquin, the Mississippi, the Orinoco, and the Paraná.

The delta of the St. Clair River between the Canadian province of Ontario and the U.S. state of Michigan is the largest delta emptying into a body of fresh water.

Ecological Threats to Deltas

Human activities, including diversion of water and the creation of dams for hydroelectric power or to create reservoirs can radically alter delta ecosystems. Dams block sedimentation which can cause the delta to erode away. The use of water upstream can greatly increase salinity levels as less fresh water flows to meet the salty ocean water. While nearly all deltas have been impacted to some degree by humans, the Nile Delta and Colorado River Delta are some of the most extreme examples of the ecological devastation caused to deltas by damming and diversion of water.

Deltas in the Economy

Ancient deltas are a benefit to the economy due to their well sorted sand and gravel. Sand and gravel is often quarried from these old deltas and used in concrete for highways, buildings, sidewalks, and even landscaping. More than 1 billion tons of sand and gravel are produced in the United States alone. Not all sand and gravel quarries are former deltas, but for ones that are, a lot of the sorting is already done by the power of water.

As lowlands often adjacent to urban areas, deltas often comprise extensive industrial and commercial areas as well as agricultural land. These uses are often in conflict. The Fraser Delta in British Columbia, Canada, includes the Vancouver Airport and the Roberts Bank Superport and the Annacis Island industrial zone, and a mix of commercial, residential and agricultural land. Space is so limited in the Lower Mainland region and in British Columbia in general, which is very mountainouse, that in order to preserve agricultural land for food production, the Agricultural Land Reserve was created.

Deltas on Mars

Researchers have found a number of examples of deltas that formed in Martian lakes. Finding deltas is a major sign that Mars once had a lot of water. Deltas have been found over a wide geographical range. Below are pictures of a few.

Lagoon

A lagoon is a shallow body of water separated from a larger body of water by barrier islands or reefs. Lagoons are commonly divided into coastal lagoons and atoll lagoons. They have also been identified as occurring on mixed-sand and gravel coastlines. There is an overlap between bodies of water classified as coastal lagoons and bodies of water classified as estuaries. Lagoons are common coastal features around the world.

Definition

Figure: *Garabogaz-Göl lagoon in Turkmenistan*

Lagoons are shallow, often elongated bodies of water separated from a larger body of water by a shallow or exposed shoal, coral reef, or similar feature. Some authorities (such as Nybakken) include fresh water bodies in the definition of "lagoon", while others explicitly restrict "lagoon" to bodies of water with some degree of salinity.

The distinction between "lagoon" and "estuary" also varies between authorities. Davis restricts "lagoon" to bodies of water with little or no fresh water inflow, and little or no tidal flow, and calls any bay that receives a regular flow of fresh water an "estuary". Davis does state that the terms "lagoon" and "estuary" are "often loosely applied, even in scientific literature." Kusky characterizes lagoons as normally being elongated parallel to the coast, while estuaries are usually drowned river valleys, elongated perpendicular to the coast. When used within the context of a distinctive portion of coral reef ecosystems, the term "lagoon" is synonymous with the term "back reef" or "backreef", which is more commonly used by coral reef scientists to refer to the same area. Coastal lagoons are classified as inland bodies of water.

Many lagoons do not include "lagoon" in their common names. Albemarle and Pamlico sounds in North Carolina, Great South Bay between Long Island and the barrier beaches of Fire Island in New York, Isle of Wight Bay, which separates Ocean City, Maryland from the rest of Worcester County, Maryland, Banana River in Florida, Lake Illawarra in New South Wales, Montrose Basin in Scotland, and Broad Water in Wales have all been classified as lagoons, despite their names. In England, The Fleet at Chesil Beach has also been described as a lagoon.

In Latin America, the term "laguna", which lagoon translates to, is often used to describe a lake, such as Laguna Catemaco. In Portuguese, "lagoa" may be a body of shallow sea water, but also a relatively small freshwater lake not linked to the sea.

Etymology

"Lagoon" is derived from the Italian *laguna*, which refers to the waters around Venice, the Lagoon of Venice. *Laguna* is attested in English by at least 1612, and had been Anglicized to "lagune" by 1673. In 1697 William Dampier referred to a "Lagune or Lake of Salt water" on the coast of Mexico. Captain James Cook described an island "of Oval form with a Lagoon in the middle" in 1769.

Atoll Lagoons

Atoll lagoons form as coral reefs grow upwards while the islands that the reefs surround subside, until eventually only the reefs remain above sea level. Unlike the lagoons that form shoreward of fringing reefs, atoll lagoons often contain some deep (>20m) portions.

Coastal Lagoons

Coastal lagoons form along gently sloping coasts where barrier islands or reefs can develop off-shore, and the sea-level is rising relative to the land along the shore (either because of an intrinsic rise in sea-level, or subsidence of the land along the coast). Coastal lagoons do not form along steep or rocky coasts, or if the range of tides is more than 4 metres (13 ft). Due to the gentle slope of the coast, coastal lagoons are shallow. They are sensitive to changes in sea level. A relative drop in sea level may leave a lagoon largely dry, while a rise in sea level may let the sea breach or destroy barrier islands, and leave reefs too deep under water to protect the lagoon. Nybakken describes coastal lagoons and barrier islands as a "coupled system".

Coastal lagoons are young and dynamic, and may be short-lived in geological terms. Coastal lagoons are common, occurring along nearly 15 percent of the world's shorelines. In the United States, lagoons are found along more than 75 percent of the eastern and Gulf coasts.

Coastal lagoons are usually connected to the open ocean by inlets between barrier islands. The number and size of the inlets, precipitation, evaporation, and inflow of fresh water all affect the nature of the lagoon. Lagoons with little or no interchange with the open ocean, little or no inflow of fresh water, and high evaporation rates, such as Lake St. Lucia, in South Africa, may become highly saline. Lagoons with no connection to the open ocean and significant inflow of fresh water, such as the Lake Worth Lagoon in Florida in the middle of the 19th century, may be entirely fresh.

On the other hand, lagoons with many wide inlets, such as the Wadden Sea, have strong tidal currents and mixing. Coastal lagoons tend to accumulate sediments from inflowing rivers, from runoff from the shores of the lagoon, and from sediment carried into the lagoon through inlets by the tide. Large quantities of sediment may be occasionally be deposited in a lagoon when storm waves overwash barrier islands. Mangroves and marsh plants can facilitate the accumulation of sediment in a lagoon. Benthic organisms may stabilize or destabilize sediments.

River-mouth Lagoons on Mixed Sand and Gravel Beaches

River-mouth lagoons on mixed sand and gravel (MSG) beaches form at the river-coast interface where a typically braided, although sometimes meandering, river interacts with a coastal environment that is significantly affected by longshore drift. The lagoons which form on the MSG coastlines are common on the east coast of the South Island, New Zealand and have long been referred to as 'hapua' by the Mâori. This classification differentiates hapua from similar lagoons located on the New Zealand coast termed 'waituna'.

Hapua are often located on paraglacial coastal areas where there is a low level of coastal development and minimal population density. Hapua form as the river carves out an elongated coast-parallel area, blocked from the sea by a MSG barrier which constantly alters its shape and volume due to longshore drift. Longshore drift continually extends the barrier behind which the hapua forms by transporting sediment along the coast. Hapua are defined as a narrow shore-parallel extensions of the coastal riverbed. They discharge the majority of stored water to the ocean via an ephemeral and highly mobile drainage channel or outlet. The remainder percolates through the MSG barrier due to its high levels of permeability. Hapua systems are driven by a wide range of dynamic processes that are generally classified as fluvial or marine; changes in the balance between these processes as well as the antecedent barrier conditions can cause shifts in the morphology of the hapua, in particular the barrier. New Zealand examples include the Rakaia, Ashburton and Hurunui river-mouths.

Hapua Environment

Hapua have been identified as establishing in the Canterbury Bight coastal region on the east coast of the South Island. They are often found in areas of coarse-grained sediment where contributing rivers have moderately steep bed gradients. MSG beaches in the Canterbury Bight region contain a wide range of sediment sizes from sand to boulders and are exposed to the high energy waves that make up an east coast swell environment. MSG beaches are reflective rather than dissipative energy zones due to their morphological characteristics. They have a steep foreshore which is known as the 'engine room' of the beach profile. In this zone, swash and backwash are dominating processes alongside longshore transport. MSG beaches do not have a surf zone; instead a single line of breakers is visible in all sea conditions. Hapua are associated with MSG beaches as the variation in sediment size allows for the barrier to be permeable.

The east coast of the South Island has been identified as being in a period of chronic erosion of approximately 0.5 metres per year. This erosion trend is a result of a number of factors. According to the classification scheme of Zenkovich, the rivers on the east coast can be described as 'small'; this classification is not related to their flow rate but to the insufficient amount of sediment that they transport to the coast to nourish it. The sediment provided is not adequate to nourish the coast against its typical high energy waves and strong longshore drift. These two processes constantly remove sediment depositing it either offshore or further up drift. As the coastline becomes eroded the hapua have been 'rolling back' by eroding the backshore to move landwards.

Hapua or river-mouth lagoons form in micro-tidal environments. A micro-tidal environment is where the tidal range (distance between low tide and high tide) is less than two metres. Tidal currents in a micro-tidal zone are less than those found on meso-tidal (two – four metres) and macro-tidal (greater than four metres) coastlines. Hapua form in this type of tidal environment as the tidal currents are unable to compete with the powerful freshwater flows of the rivers therefore there is no negligible tidal penetration to the lagoon. A fourth element of the environment in which hapua form is the strong longshore drift component. Longshore or littoral drift is the transportation of sediments along the coast at an angle to the shoreline.

In the Canterbury Bight coastal area; the dominant swell direction is northwards from the Southern Ocean. Therefore the principal movement of sediment via longshore drift is north towards Banks Peninsula. Hapua are located in areas dominated by longshore drift; because it aids the formation of the barrier behind which the hapua is sited.

A hapua also requires sediment to form the lagoon barrier. Sediment which nourishes the east coast of New Zealand can be sourced from three different areas. Material from the highly erodible Southern Alps is removed via weathering; then carried across the Canterbury Plains by various braided rivers to the east coast beaches. The second source of sediment is the high cliffs which are located in the hinterland of lagoons. These can be eroded during the occurrence of high river flow or sea storm events. Beaches further south provide nourishment to the northern coast via longshore transport.

Hapua Characteristics

Hapua have a number of characteristics which includes shifts between a variety of morphodynamic states due to changes in the balance between marine and fluvial processes as well as the antecedent barrier conditions. The MSG barrier constantly changes size and shape as a result of the longshore drift. Water stored in the hapua drains to the coast predominately though an outlet; although it can also seep through the barrier depending on the permeability of the material.

Changes in the level of the lagoon water do not occur as a result of saltwater or tidal intrusion. Water in a hapua is predominately freshwater originating from the associated river. Hapua are non-estuarine, there is no tidal inflow however the tide does have an effect on the level of water in the lagoon. As the tide reaches its peak, the lagoon water has a much smaller amount of barrier to permeate through so the lagoon level rises. This is related to a physics theory known as hydraulic head. The lagoon level has a similar sinusoidal wave shape as the tide but reaches its peak slightly later. In general, any saltwater intrusion into the hapua will only occur during a storm via wave overtopping or sea spray.

Hapua can act as both a source and sink of sediment. The majority of sediment in the hapua is fluvial sourced. During medium to low river flows, coarser sediment generally collects in the hapua; while some of the finer sediment can be transported through the outlet to the coast. During flood events the hapua is 'flushed out' with larger amounts of sediment transferred through the outlet. This sediment can be deposited offshore or downdrift of the hapua replenishing the undernourished beach. If a large amount of material is released to the coast at one time it can be identified as a 'slug'. These can often be visible from aerial photographs.

Antecedent barrier conditions combined with changes in the balance between marine and fluvial processes results in shifts between a variety of morphological states in a hapua or river-mouth lagoon on a MSG beach. Marine processes includes the direction of wave approach, wave height and the coincidence of storm waves with high tides. Marine processes tend to dominate the majority of morphodynamic conditions until there is a large enough flood event in the associated river to breach the barrier. The level and frequency of base or flood flows are attributed to fluvial processes. Antecedent barrier conditions are the permeability, volume and height of the barrier as well as the width and presence of previous outlet channels. During low to medium river flows, the outlet from the lagoon to the sea becomes offset in the direction of longshore drift. Outlet efficiency tends to decrease the further away from the main river-mouth the outlet is. A decrease in efficiency can cause the outlet to become choked with sediment and the hapua to close temporarily. The potential for closure varies between different hapua depending on whether marine or fluvial processes are the bigger driver in the event. A high flow event; such as a fresh or flood can breach the barrier directly opposite the main river channel. This causes an immediate decrease in the water level of the hapua; as well as transporting previously deposited sediments into the ocean. Flood events are important for eroding lagoon back shores; this is a behaviour which allows hapua to retreat landward and thus remain coastal landforms even with coastal transgression and sea level rise. During high flow events there is also the possibility for secondary breaches of the barrier or lagoon truncation to occur.

Storm events also have the ability to close hapua outlets as waves overtop the barrier depositing sediment and choking the scoured channel. The resultant swift increase in lagoon water level causes a new outlet to be breached rapidly due to the large hydraulic head that forms between the lagoon and sea water levels. Storm breaching is believed to be an important but unpredictable control on the duration of closures at low to moderate river flow levels in smaller hapua.

Hapua are extremely important for a number of reasons. They provide a link between the river and sea for migrating fish as well as a corridor for migratory birds. To lose this link via closure of the hapua outlet could result in losing entire generations of specific species as they may need to migrate to the ocean or the river as a vital part of their lifecycle. River-mouth lagoons such as hapua were also used a source for *mahinga kai* (food gathering) by the Mâori people. However, this is no longer the case due to catchment degradation which has resulted in lagoon deterioration. River-mouth lagoons on MSG beaches are not well explained in international literature.

Hapua Case Study

The hapua located at the mouth of the Rakaia River stretches approximately three kilometres north from where the river-mouth reaches the coast. The average width of the hapua between 1952 and 2004 was approximately 50 metres; whilst the surface area has stabilised at approximately 600,000 square metres since 1966. The coastal hinterland is composed of erodible cliffs and a low lying area commonly known as the Rakaia Huts. This area has changed notably since European Settlement; with the drainage of ecologically significant wetlands and development of the small bach community.

The Rakaia River begins in the Southern Alps, providing approximately 4.2 Mt per year of sediment to the east coast. It is a braided river with a catchment area of 3105 kilometres squared and a mean flow of 221 cubic metres per second. The mouth of the Rakaia River reaches the coast south of Banks Peninsula. As the river reaches the coast it diverges into two channels; with the main channel flowing to the south of the island. As the hapua is located in the Canterbury Bight it is in a state of constant morphological change due to the prevailing southerly sea swells and resultant northwards longshore drift.

2

Coastal Features Formed by Sediment

Beach

A beach is a landform along the shoreline of an ocean, sea, lake, or river. It usually consists of loose particles, which are often composed of rock, such as sand, gravel, shingle, pebbles, or cobblestones. The particles comprising the beach are occasionally biological in origin, such as mollusc shells or coralline algae.

Wild beaches are beaches that do not have lifeguards or trappings of modernity nearby, such as resorts, camps, and hotels. They are sometimes called undeclared, undeveloped, or undiscovered beaches. Wild beaches can be valued for their untouched beauty and preserved nature. They are most commonly found in less developed areas including, for example, parts of Puerto Rico, the Dominican Republic, Thailand, the Philippines and Indonesia, but they are also found in developed nations such as Australia and New Zealand.

Beaches typically occur in areas along the coast where wave or current action deposits and reworks sediments.

Overview

1. Swash zone: is alternately covered and exposed by wave run-up.
2. Beach face: sloping section below berm that is exposed to the swash of the waves.
3. Wrack line: the highest reach of the daily tide where organic and inorganic debris is deposited by wave action.
4. Berm: Nearly horizontal portion that stays dry except during extremely high tides and storms. May have sand dunes.

Although the seashore is most commonly associated with the word *beach*, beaches are found by lakes and alongside large rivers.

Beach may refer to:

- small systems where rock material moves onshore, offshore, or alongshore by the forces of waves and currents; or
- geological units of considerable size.

The former are described in detail below; the larger geological units are discussed elsewhere under bars.

There are several conspicuous parts to a beach that relate to the processes that form and shape it. The part mostly above water (depending upon tide), and more or less actively influenced by the waves at some point in the tide, is termed the beach berm. The berm is the deposit of material comprising the active shoreline. The berm has a *crest* (top) and a *face* — the latter being the slope leading down towards the water from the crest. At the very bottom of the face, there may be a *trough*, and further seaward one or more long shore bars: slightly raised, underwater embankments formed where the waves first start to break.

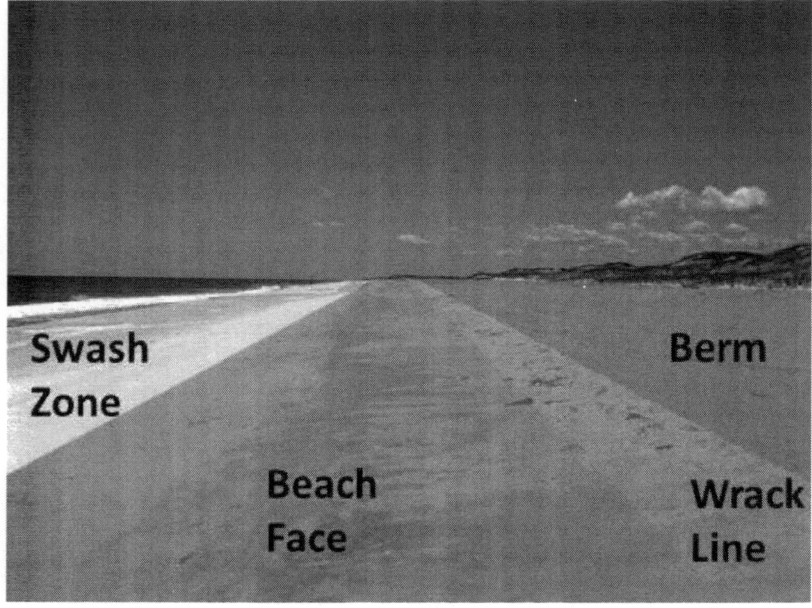

Figure: *The four sections of most beaches.*

The sand deposit may extend well inland from the *berm crest*, where there may be evidence of one or more older crests (the *storm beach*) resulting from very large storm waves and beyond the influence of the normal waves. At some point the influence of the waves (even storm waves) on the material comprising the beach stops, and if the particles are small enough (sand size or smaller), winds shape the feature. Where wind is the force distributing the grains inland, the deposit behind the beach becomes a *dune*.

These geomorphic features compose what is called the *beach profile*. The beach profile changes seasonally due to the change in wave energy experienced during summer and winter months. In temperate areas where summer is characterised by calmer seas and longer periods between breaking wave crests, the beach profile is higher in summer. The gentle wave action during this season tends to transport sediment up the beach towards the berm where it is deposited and remains while the water recedes. Onshore winds carry it further inland forming and enhancing dunes.

Conversely, the beach profile is lower in the storm season (winter in temperate areas) due to the increased wave energy, and the shorter periods between breaking wave crests. Higher energy waves breaking in quick succession tend to mobilise sediment from the shallows, keeping it in suspension where it is prone to be carried along the beach by longshore currents, or carried out to sea to form longshore bars, especially if the longshore current meets an outflow from a river or flooding stream. The removal of sediment from the beach berm and dune thus decreases the beach profile.

In tropical areas, the storm season tends to be during the summer months, with calmer weather commonly associated with the winter season.

If storms coincide with unusually high tides, or with a freak wave event such as a tidal surge or tsunami which causes significant coastal flooding, substantial quantities of material may be eroded from the coastal plain or dunes behind the berm by receding water. This flow may alter the shape of the coastline, enlarge the mouths of rivers and create new deltas at the mouths of streams that formerly were not powerful enough to overcome longshore movement of sediment.

The line between beach and dune is difficult to define in the field. Over any significant period of time, sediment is always being exchanged between them. The *drift line* (the high point of material deposited by waves) is one potential demarcation. This would be the point at which significant wind movement of sand could occur, since the normal waves do not wet the sand beyond this area. However, the drift line is likely to move inland under assault by storm waves.

Beach Formation

Beaches are the result of wave action by which waves or currents move sand or other loose sediments of which the beach is made as these particles are held in suspension. Alternatively, sand may be moved by saltation (a bouncing movement of large particles). Beach materials come from erosion of rocks offshore, as well as from headland erosion and slumping producing deposits of scree. Some of the whitest sand in the world, along Florida's Emerald Coast, comes from the erosion of quartz in the Appalachian Mountains.

A coral reef offshore is a significant source of sand particles. Some species of fish that feed on algae attached to coral outcrops and rocks can create substantial quantities of sand particles over their lifetime as they nibble during feeding, digesting the organic matter, and discarding the rock and coral particles which pass through their digestive tracts.

The composition of the beach depends upon the nature and quantity of sediments upstream of the beach, and the speed of flow and turbidity of water and wind. Sediments are moved by moving water and wind according to their particle size and state of compaction. Particles tend to settle and compact in still water. Once compacted, they are more resistant to erosion. Established vegetation (especially species with complex network root systems) will resist erosion by slowing the fluid flow at the surface layer.

When affected by moving water or wind, particles that are eroded and held in suspension will increase the erosive power of the fluid that holds them by increasing the average density, viscosity and volume of the moving fluid.

The nature of sediments found on a beach tends to indicate the energy of the waves and wind in the locality. Coastlines facing very energetic wind and wave systems will tend to hold only large rocks as smaller particles will be held in suspension in the turbid water column and carried to calmer areas by longshore currents and tides. Coastlines that are protected from waves and winds will tend to allow finer sediments such as clays and mud to precipitate creating mud flats and mangrove forests.

The shape of a beach depends on whether the waves are constructive or destructive, and whether the material is sand or shingle. Waves are constructive if the period between their wave crests is long enough for the breaking water to recede and the sediment to settle before the succeeding wave arrives and breaks. Fine sediment transported from lower down the beach profile will compact if the receding water percolates or soaks into the beach. Compacted sediment is more resistant to movement by turbulent water from succeeding waves.

Conversely, waves are destructive if the period between the wave crests is short. Sediment that remains in suspension when the following wave crest arrives will not be able to settle and compact and will be more susceptible to erosion by longshore currents and receding tides.

Constructive waves move material up the beach while destructive waves move the material down the beach. During seasons when destructive waves are prevalent, the shallows will carry an increased load of sediment and organic matter in suspension. On sandy beaches, the turbulent backwash of destructive waves removes material forming a gently sloping beach. On pebble and shingle beaches the swash is dissipated more quickly because the large particle size allows greater percolation, thereby reducing the power of the backwash, and the beach remains steep.

Compacted fine sediments will form a smooth beach surface that resists wind and water erosion. During hot calm seasons, a crust may form on the surface of ocean beaches as the heat of the sun evaporates the water leaving the salt which crystallises around the sand particles. This crust forms an additional protective layer that resists wind erosion unless disturbed by animals, or dissolved by the advancing tide. Cusps and horns form where incoming waves divide, depositing sand as horns and scouring out sand to form cusps. This forms the uneven face on some sand shorelines.

Beach Erosion and Accretion

Natural Erosion and Accretion

Causes: Beaches are changed in shape chiefly by the movement of water and wind. Any weather event that is associated with turbid or fast flowing water, or high winds will erode exposed beaches. Longshore currents will tend to replenish beach sediments and repair storm damage. Tidal waterways generally change the shape of their adjacent beaches by small degrees with every tidal cycle. Over time these changes can become substantial leading to significant changes in the size and location of the beach.

Effects on Flora

Changes in the shape of the beach may undermine the roots of large trees and other flora. Many beach adapted species (such as coconut palms) have a fine root system and large root ball which tends to withstand wave and wind action and tends to stabilize beaches better than other trees with a lesser root ball.

Effects on Adjacent Land

Erosion of beaches can expose less resilient soils and rocks to wind and wave action leading to undermining of coastal headlands eventually resulting in catastrophic collapse of large quantities of overburden into the shallows. This material may be distributed along the beach front leading to a change in the habitat as sea grasses and corals in the shallows may be buried or deprived of light and nutrients.

Manmade Erosion and Accretion

Coastal areas settled by man inevitably become subject to the effects of man-made structures and processes. Over long periods of time these influences may substantially alter the shape of the coastline, and the character of the beach.

Destruction of Flora

Beach front flora plays a major role in stabilizing the foredunes and preventing beach head erosion and inland movement of dunes. If flora with network root systems (creepers, grasses and palms) are able to become established, they provide an effective coastal defence as they trap sand particles and rainwater and enrich the surface layer of the dunes, allowing other plant species to become established. They also protect the berm from erosion by high winds, freak waves and subsiding flood waters.

Over long periods of time, well stabilized foreshore areas will tend to accrete, while unstabilized foreshores will tend to erode, leading to substantial changes in the shape of the coastline. These changes usually occur over periods of many years. Freak wave events such as tsunami, tidal waves, and storm surges may substantially alter the shape, profile and location of a beach within hours.

Destruction of flora on the berm by the use of herbicides, excessive pedestrian or vehicular traffic, or disruption to fresh water flows may lead to erosion of the berm and dunes. While the destruction of flora may be a gradual process that is imperceptible to regular beach users, it often becomes immediately apparent after storms associated with high winds and freak wave events that can rapidly move large volumes of exposed and unstable sand, depositing them further inland, or carrying them out into the permanent water forming offshore bars, lagoons or increasing the area of the beach exposed at low tide.

Large and rapid movements of exposed sand can bury and smother flora in adjacent areas, aggravating the loss of habitat for fauna, and enlarging the area of instability. If there is an adequate supply of sand, and weather conditions do not allow vegetation to recover and stabilize the sediment, wind-blown sand can continue to advance, engulfing and permanently altering downwind landscapes.

Sediment moved by waves or receding flood waters can be deposited in coastal shallows, engulfing reed beds and changing the character of underwater flora and fauna in the coastal shallows.

Burning or clearance of vegetation on the land adjacent to the beach head, for farming and residential development, changes the surface wind patterns, and exposes the surface of the beach to wind erosion.

Farming and residential development are also commonly associated with changes in local surface water flows. If these flows are concentrated in storm water drains emptying onto the beach head, they may erode the beach creating a lagoon or delta.

Dense vegetation tends to absorb rainfall reducing the speed of runoff and releasing it over longer periods of time. Destruction by burning or clearance of the natural vegetation tends to increase the speed and erosive power of runoff from rainfall. This runoff will tend to carry more silt and organic matter from the land onto the beach and into the sea. If the flow is constant, runoff from cleared land arriving at the beach head will tend to deposit this material into the sand changing its colour, odor and fauna.

Creation of Beach Access Points

The concentration of pedestrian and vehicular traffic accessing the beach for recreational purposes may cause increased erosion at the access points if measures are not taken to stabilize the beach surface above high-water mark. Recognition of the dangers of loss of beach front flora has caused many local authorities responsible for managing coastal areas to restrict beach access points by physical structures or legal sanctions, and fence off foredunes in an effort to protect the flora. These measures are often associated with the construction of structures at these access points to allow traffic to pass over or through the dunes without causing further damage.

Concentration of Runoff

Beaches provide a filter for runoff from the coastal plain. If the runoff is naturally dispersed along the beach, water borne silt and organic matter will be retained on the land and will feed the flora in the coastal area. Runoff that is dispersed along the beach will tend to percolate through the beach and may emerge from the beach at low tide.

The retention of the fresh water may also help to maintain underground water reserves and will resist salt water incursion. If the surface flow of the runoff is diverted and concentrated by drains that create constant flows over the beach above the sea or river level, the beach will be eroded and ultimately form an inlet unless longshore flows deposit sediments to repair the breach.

Once eroded, an inlet may allow tidal inflows of salt water to pollute areas inland from the beach and may also affect the quality of underground water supplies and the height of the water table.

Deprivation of Runoff

Some flora naturally occurring on the beach head requires fresh water runoff from the land. Diversion of fresh water runoff into drains may deprive these plants of their water supplies and allow sea water incursion, increasing the saltiness of the ground water. Species that are not able to survive in salt water may die and be replaced by mangroves or other species adapted to salty environments.

Inappropriate Beach Nourishment

Beach nourishment is the importing and deposition of sand or other sediments in an effort to restore a beach that has been damaged by erosion. Beach nourishment often involves excavation of sediments from riverbeds or sand quarries. This excavated sediment may be substantially different in size and appearance to the naturally occurring beach sand.

In extreme cases, beach nourishment may involve placement of large pebbles or rocks in an effort to permanently restore a shoreline subject to constant erosion and loss of foreshore. This is often required where the flow of new sediment caused by the longshore current has been disrupted by construction of harbours, breakwaters, causeways or boat ramps, creating new current flows that scour the sand from behind these structures, and deprive the beach of restorative sediments. If the causes of the erosion are not addressed, beach nourishment can become a necessary and permanent feature of beach maintenance.

During beach nourishment activities, care must be taken to place new sediments so that the new sediments compact and stabilize before aggressive wave or wind action can erode them. Material that is concentrated too far down the beach may form a temporary groyne that will encourage scouring behind it. Sediments that are too fine or too light may be eroded before they have compacted or been integrated into the established vegetation. Foreign unwashed sediments may introduce flora or fauna that are not usually found in that locality.

Brighton Beach, on the south coast of England, is a shingle beach that has been nourished with very large pebbles in an effort to withstand erosion of the upper area of the beach. These large pebbles made the beach unwelcoming for pedestrians for a period of time until natural processes integrated the naturally occurring shingle into the pebble base.

Beach Access Design

Beach access is an important consideration where substantial numbers of pedestrians or vehicles require access to the beach. Allowing random access across delicate foredunes is seldom considered good practice as it is likely to lead to destruction of flora and consequent erosion of the fore dunes.

A well designed Beach Access should:-

- provide a durable surface able to withstand the traffic flow;
- aesthetically complement the surrounding structures and natural landforms;
- be located in an area that is convenient for users and consistent with safe traffic flows;
- be scaled to match the traffic flow (i.e. wide and strong enough to safely carry the size and quantity of pedestrians and vehicles intended to use it);
- be maintained appropriately; and
- be signed and lit to discourage beach users from creating their own alternative crossings that may be more destructive to the beachhead.

Concrete Ramp or Steps

A concrete ramp should follow the natural profile of the beach to prevent it from changing the normal flow of waves, longshore currents, water and wind. A ramp that is below the beach profile will tend to become buried and cease to provide a good surface for vehicular traffic. A ramp or stair that protrudes above the beach profile will tend to disrupt longshore currents creating deposits in front of the ramp, and scouring behind. Concrete ramps are the most expensive vehicular beach accesses to construct requiring use of a quick drying concrete or a coffer dam to protect them from tidal water during the concrete curing process. Concrete is favoured where traffic flows are heavy and access is required by vehicles that are not adapted to soft sand (e.g. road registered passenger vehicles and boat trailers).

Concrete stairs are commonly favoured on beaches adjacent to population centres where beach users may arrive on the beach in street shoes, or where the foreshore roadway is substantially higher than the beach head and a ramp would be too steep for safe use by pedestrians. A composite stair ramp may incorporate a central or side stair with one or more ramps allowing pedestrians to lead buggies or small boat dollies onto the beach without the aid of a powered vehicle or winch. Concrete ramps and steps should be maintained to prevent buildup of moss or algae that may make their wet surfaces slippery and dangerous to pedestrians and vehicles.

Corduroy (Beach Ladder)

A corduroy or beach ladder (or board and chain) is an array of planks (usually hardwood or treated timber) laid close together and perpendicular to the direction of traffic flow, and secured at each end by a chain or cable to form a pathway or ramp over the sand dune. Corduroys are cheap and easy to construct and quick to deploy or relocate. They are commonly used for pedestrian access paths and light duty vehicular access ways. They naturally conform to the shape of the underlying beach or dune profile, and adjust well to moderate erosion, especially longshore drift.

However, they can cease to be an effective access surface if they become buried or undermined by erosion by surface runoff coming from the beach head. If the corduroy is not wide enough for vehicles using it, the sediment on either side may be displaced creating a spoon drain that accelerates surface run off and can quickly lead to serious erosion. Significant erosion of the sediment beside and under the corduroy can render it completely ineffective and make it dangerous to pedestrian users who may fall between the planks.

Fabric Ramp

Fabric ramps are commonly employed by the military for temporary purposes where the underlying sediment is stable and hard enough to support the weight of the traffic. A sheet of porous fabric is laid over the sand to stabilize the surface and prevent vehicles from bogging. Fabric Ramps usually cease to be useful after one tidal cycle as they are easily washed away, or buried in sediment.

Foliage Ramp

A foliage ramp is formed by planting resilient species of hardy plants such as grasses over a well formed sediment ramp. The plants may be supported while they become established by placement of layers of mesh, netting, or coarse organic material such as vines or branches. This type of ramp is ideally suited for intermittent use by vehicles with a low wheel loading such as dune buggies or agricultural vehicles with large tyres. A foliage ramp should require minimal maintenance if initially formed to follow the beach profile, and not overused.

Gravel Ramp

A gravel ramp is formed by excavating the underlying loose sediment and filling the excavation with layers of gravel of graduated sizes as defined by John Loudon McAdam.

The gravel is compacted to form a solid surface according to the needs of the traffic. Gravel ramps are less expensive to construct than concrete ramps and are able to carry heavy road traffic provided the excavation is deep enough to reach solid subsoil. Gravel ramps are subject to erosion by water. If the edges are retained with boards or walls and the profile matches the surrounding beach profile, a gravel ramp may become more stable as finer sediments are deposited by percolating water.

Beach Cusps

Beach cusps are shoreline formations made up of various grades of sediment in an arc pattern. The horns are made up of coarser materials and the embayment contains all the finer grain sediment. They can be found all over the world and are most noticeable on shorelines with coarser sediment such as pebble beaches, however they can occur with sediment of any size. They nearly always occur in a regular pattern with cusps of equal size and spacing appearing along stretches of the shoreline.

These cusps are most often a few metres long, however they may reach 60 m (200 ft) across. Although the origin of beach cusps has yet to be proven, once cusps have been created they are a self-sustaining formation. This is because when an oncoming wave hits the horn of a beach cusp, it is split and forced into two directions.

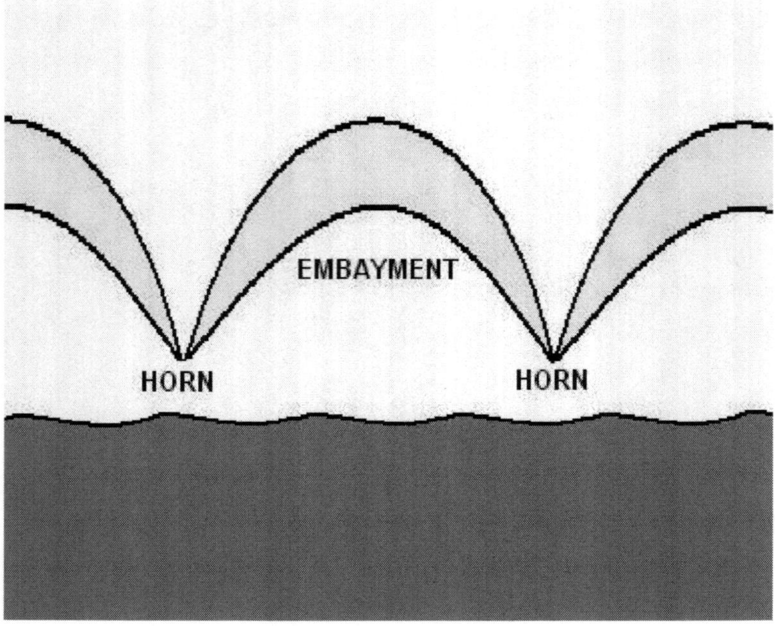

Figure: *A beach cusp*

The crashing of the wave into the cusps slows its velocity, causing coarser sediment to fall out of suspension and be deposited on the horns. The waves then flow along the embayments (picking up finer sediment) and run into one another in the middle. After this collision these waves attempt to flow back out to sea where they are met by incoming waves. Therefore, once the cusp is established, coarser sediment is constantly being deposited on the horn and finer sediment is being eroded away from the embayments. This process causes the horns and embayments to at least maintain their size, if not grow larger.

Formation

There are a number of theories as to why beach cusps are formed but currently, there are only two explanations with any real credibility.

Standing Edge Wave Theory

The standing edge wave theory is based on an interaction, near the shoreline, between the waves that are approaching the shore and waves that have been set up perpendicular to the shoreline called 'edge waves'. The regular arrival of incoming waves in the near shore waters causes the development of waves perpendicular to the direction of the incoming waves; these are termed 'edge waves'. These edge waves become trapped near the shoreline and when two of them come together from opposite directions, a standing edge wave is formed. The movement patterns of these waves are fixed and so can be defined as two regions of interest, the nodal and antinodal points.

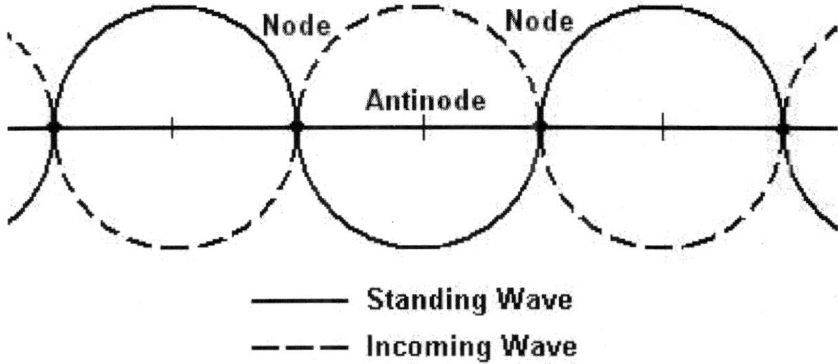

Figure: *Standing Edge Wave Theory*

The antinodal point is where all the movement takes place as the water rises and falls, creating a series of peaks and troughs. Between these antinodal points are the nodal points where no vertical movement takes place. An incoming wave has an almost uniform height but when it collides with a standing edge wave, this is changed. If it collides with a peak, then the wave height is increased and if it collides with a trough, then its height is decreased. In the diagram on the right, the two waves cancel each other out creating a flat surface however this is a highly simplified version of events. The incoming wave has the same wave period as the edge wave, so the incoming wave changes from a peak to a trough over the same period as it takes the standing wave to change so they keep the same pattern. These are known as synchronous waves and are very uncommon.

The more common standing edge waves are subharmonic and these can have a wave period twice that of the incoming wave. This produces a far more complex system of waves as by the time the incoming wave has completed one cycle from peak to trough, the standing edge waves have done two. So what started as the peak of a standing edge wave within the trough of the incoming wave will change to a trough before the incoming wave has changed so what was initially being given a boost in height now experiences a drop. Essentially what this means is that there are a regularly spaced series of peaks and troughs along the length on the incoming wave that are caused by its interaction with the standing edge waves and it is these that caused the development of beach cusps.

In areas where the wave height has been increased, the wave now has more power and so can erode more and in areas where the wave height has been decreased, the wave now has less power and so will not erode as much. This is what forms the cusps as the areas with high erosion become the embayments of the cusp and the areas with low erosion become the horns.

The problem with the standing edge wave theory is that it would only account for the initial formation of the cusp and not their continued growth afterwards because as the cusp increases in size the amplitude of the edge wave decreases to the point where it is no longer a factor.

Self-organization

This theory was initially dismissed due to complications with it but by using more advanced computer simulations, it has been developed to the point where it now provides a reasonable alternative to the standing edge wave theory. The theory has two main points that seek to explain the formation of regularly spaced beach cusps.

The first is that positive feedback between the morphology of the beach and the flow of the water creates relief patterns. On a flat beach, surface areas will develop with a slightly lower relief than their surroundings. Areas with lower relief attract and accelerate water particles, which means they have more energy and the area is eroded further. Through positive feedback, the area becomes more and more eroded, and this creates the embayment. The development of these areas of lower relief causes mean beach level to be lowered and so areas that were initially on the level are now below the average and as such, are now areas of higher relief. Areas of higher relief slow the water down and sediment is deposited on top of them, which increases their impact. This creates the horns.

The uniform spacing of cusps is caused by the communication of surface gradients along the beach by the smoothing of the beach surface as the beach tries to rearrange itself to reduce variations in the plane.

The second main point about the regular spacing of cusps is that negative feedback will decrease the amount of net erosion and deposition within a well-formed cusp. As the wave strikes the beach, it will first come into contact with the cusp horns, which will slow the water down. This causes it to lose energy and some of the heavier sediment that it is carrying will be deposited. The loss of this sediment, however, gives the water extra energy and it uses this to remove sediment from the embayment on the backwash.

The problem with this theory is that this method of cusp formation would take time and if you were observing their formation, then you would see a number of random cusps form along the beach, which then slowly spread along the shore as they even out in size, with small cusps joining together and larger cusps being separated in two. But in the field, cusps form a regular pattern almost instantly and they all appear at the same time.

Field Studies

In recent years, there has been considerable debate about whether beach cusp formation is associated with the presence of standing edge waves (standing edge wave theory), results from self-organizing feedback between changing topography and swash motion (self-organization theory) or is attributable to a number of other less popular mechanisms.

The National Centre of Scientific Research in France, collected a large amount of data from laboratory experiments and field studies, published over the last 50 years to test the predictions of the two main cusp forming hypotheses. These analyses, using more data than previous attempts, confirm that there is a possible link between cusp development and both edge waves and swash-sediment feedback, and that it is not possible to produce conclusive support for one theory above the other with the simple measurements that have been made previously.

The Naval Research Laboratory, the corporate research laboratory for the United States Navy and Marine Corps carried out nearly nine years of video imagery from Duck, North Carolina to determine the timing of cusp formation (to within half a day) and the distances separating consecutive cusp horns (to within half a metre). Supplementary data provided by nearshore instrumentation and surveying vehicles was used to document the environmental conditions during cusp development.

These extensive observations conclusively demonstrate that cusps at this location develop, following storms during the transition from high energy to low energy wave conditions as the wave angle approaches normal incidence. A peculiar suggestion of historisis within the cusp spacing time series was observed and may suggest that existing theories of cusp formation need to be reformulated.

Boondocks

The term boondocks is an American English term for a remote, usually brushy rural area; or to a remote city or town that is considered unsophisticated.

Origins

The expression was introduced to English by American military personnel serving in the Philippines during the early years of the 20th century. It derives from the Tagalog word *"bundok"*, meaning "mountain". According to military historian Paul Kramer, the term had attached to it "connotations of bewilderment and confusion", due to the guerrilla warfare in which the soldiers were engaged.

"Bundok" as originally used by Filipinos is a colloquialism referring to rural areas inland which are usually mountainous and difficult to access (most major Filipino cities and settlements are located near the coastline). Other equivalent terms used are the Spanish-derived *"probinsya"* ('province') and the Cebuano *"bukid"* ('mountain').

When used generally, the term refers to a rustic or uncivilized area. When referring to people (Tagalog *"taga-bundok"/"probinsyano"*, Cebuano *"taga-bukid"* – literally 'someone who comes from the mountains/provinces'), it acquires a derogatory connotation referring to the stereotype of country people being unsophisticated, ignorant, uncultured, illiterate, or naive.

Expanded Meanings

The term has evolved into American slang used to refer to *the countryside* or any implicitly isolated rural/wilderness area, regardless of topography or vegetation. Similar slang or colloquial words are "the sticks", "the wops", "the backblocks", or "Woop Woop" in Australia and New Zealand, "bundu" in South Africa, and "out in the tules" in California.

The diminutive "the boonies" can be heard in films about the Vietnam War such as Brian De Palma's *Casualties of War*. It is used by American military personnel to designate rural areas of Vietnam. Many from the urban East Coast of the United States have presumed the word to come from "boon docks," a description, not used by mariners, to describe floating docks more common in remote fishing villages than in busier ports.

"Down in the Boondocks" is a song written and produced by Joe South and sung by Billy Joe Royal. It was a hit in 1965. It tells the story of a young man who laments that people put him down because he was born in the boondocks. He is in love with the boss man's daughter and vows to work slavishly until, one day, he can "move from this old shack" and fit in her society. Throughout the song, he asks the "Lord [to] have mercy on the boy from down in the boondocks".

Cuspate Foreland

Cuspate forelands, also known as cuspate barriers or nesses in Britain, are geographical features found on coastlines and lakeshores that are created primarily by long shore drift. Formed by accretion and progradation of sand and shingle, they extend outwards from the shoreline in a triangular shape. Some cuspate forelands may be stabilised by vegetation, while others may migrate down the shoreline. Because some cuspate forelands provide an important habitat for many flora and fauna, effective management is required to reduce the impacts from both human activities and physical factors such as climate change and sea level rise.

Formation

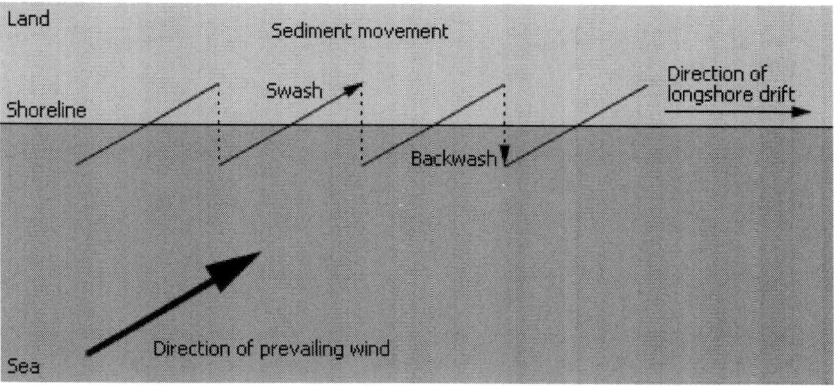

Figure: *Longshore drift is usually the main process in the formation of a cuspate foreland*

The debate involving how cuspate forelands form is ongoing. However, the most widely accepted process of formation involves long shore drift. Where longshore drift occurs in opposite directions, two spits merge into a triangular protrusion along a coastline or lakeshore. Their formation is also dependent on dominant and prevailing winds working in opposite directions. Formation can also occur when waves are diffracted around a barrier.

Cuspate forelands can form both along coastlines and along lakeshores. Those formed along coastlines can be in the lee of an offshore island, along a coastline that has no islands in the vicinity, or at a stream mouth where disposition occurs.

Formation in Narrow Straits or on Open Coastlines

A cuspate foreland can form in a strait or along a coastline that has no islands or shoals in the area. In this case, longshore drift as well as prevailing wind and waves bring sediment together from opposite directions. If there is a large angle between the waves and the shoreline, the sediment converges, accumulates, and forms beach ridges. Over time, a cuspate foreland forms as a result of continued accretion and progradation.

An example of this type of cuspate foreland is the one found at Dungeness along the southern coast of Britain. This cuspate foreland has formed as a result of the merging of SW waves from the English Channel, and waves from the east from the Strait of Dover.

Another example is the cuspate foreland found between Awatere River and White Bluffs in Marlborough, New Zealand. This foreland has ridges on the eastern and northern sides which face the prominent waves.

In other circumstances, spits are formed when long shore drift moves beach material down the beach until the coastline makes an abrupt change in direction, leading to the beach material 'spilling over' the corner to create a protrusion. This normally occurs across a river mouth.

In the case of a cuspate foreland, the prevailing wind and a powerful secondary wind in the opposite direction move shingle down the coastline from both directions to a place where the coastline changes, causing a foreland to develop. The majority of cuspate forelands are formed over a coastline that juts out into the sea at enough of an angle to allow the drifting beach material to 'spill over' as a result of long shore drift in both directions.

Formation in the Lee of an Island

A cuspate foreland can form in the lee of an island. In this case, oncoming waves are diffracted around the island, protecting the coastline from the oncoming wave fronts. Sediments brought along the shoreline via longshore drift are then able to settle and accumulate in the lee of the island where there is less wave energy.

This type of foreland has formed on the west shore of the North Island of New Zealand, in the lee of Kapiti Island. Waves refract around Kapiti Island, forming an area of low wave energy where sediment from the Waikanae River is able to settle.

There is uncertainty whether the cuspate foreland has formed as a result of sediments coming from the north via longshore drift, or whether it has formed as a result of a complex cycle of sediments moving out to the continental shelf and then back again.

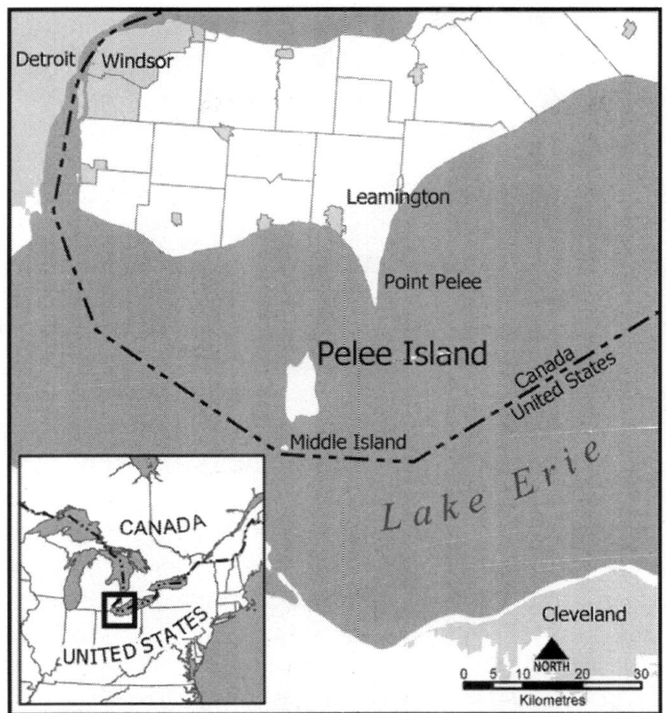

Figure: *Point Pelee cuspate foreland location*

Formation Along Lakeshores

As well as forming along coastlines, cuspate forelands can also form along lake shores, although less is known about this type of cuspate foreland. This type of cuspate foreland includes Point Pelee along the shoreline of Lake Erie, and those found along the shoreline of Lake Victoria in Australia. There are two theories with regard to the formation of Point Pelee. Firstly, it is thought that Point Pelee has formed from depositional processes. Alternatively, it is suggested that Point Pelee is a relic of a past feature that has eroded over time. This gap in knowledge provides the opportunity for further research. It is likely that Point Pelee is migrating westwards since accretion is occurring on the western side, and erosion is occurring on the eastern side. Lake Victoria in Australia also has a number of cuspate forelands. Point Scott is a cuspate foreland along this lakeshore that has formed from the gradual accumulation of sand and gravel.

Features

Cuspate forelands can be separated into three distinct areas: the central nose or apex, and two marginal wings. The apex usually has ridges that run parallel to the converging shorelines. Cuspate forelands can extend up to 5km from the shoreline, and an underwater shoal may extend much further, up to 15km from the exposed apex. Located between the mainland and the foreland are often lagoons or marshy areas. In some areas, such as along the North Carolina coastline, a series of cuspate forelands may form at least 100km apart. In areas that have a large amount of shingle, such as the cuspate foreland at Dungeness, it is also common for a fresh water table to be present.

Movement

Once formed, cuspate forelands can remain where they are and continue to develop as sediment accumulates, or alternatively they may migrate down the coast as one side of the foreland erodes and the other side accretes. Cuspate Forelands that move are typical of those that are formed on open coastlines. The direction of migration is often indicated by a series of successive beach ridges on the advancing side of the foreland where there is less wave energy. The movement of cuspate forelands is commonly explained by longshore drift acting as the main process. However, there have been observed cases where two cuspate forelands on the same shoreline have migrated in opposite directions, showing that longshore drift does not always provide a sufficient explanation for their migration.

If there is an offshore sandbank present, the position of the cuspate foreland is usually related to its position. If there is a change in the position of the sandbank, the position of the cuspate foreland typically follows. Not only does the sandbank act like an island since it causes waves to refract around it, but it also provides a source of sediment. As sand erodes from the sandbank, it is pushed towards the coastline, contributing to the formation of the cuspate foreland as the sandbank migrates along the coast. This often occurs in the opposite direction to longshore drift.

In the case of a cuspate foreland that has formed close to an island, it is possible for it to extend right up to the island, forming a tombolo. Depending on the physical conditions such as storms, the feature can alternate between a cuspate foreland and a tombolo. Gabo Island in South Australia is an example of where this occurs.

Succession

After the formation of the cuspate foreland into its distinctive triangular shape, it will start to be colonised by pioneer species that are hardy and tough enough to survive in the environment. These pioneer species secure the cuspate foreland and allow a greater amount of sediment to further secure it. Colonization and succession of vegetation is dependent on a number of factors. Firstly, if the shingle is too coarse, the amount of fine sediment that can remain between the spaces is reduced, and the likelihood that seeds will germinate and grow upwards is low. Seeds will also fail to germinate and grow if there is insufficient retention of fresh water. Stable cuspate forelands that are composed of shingle often have vegetation above the high tide line. As vegetation is established, mites and collembolans break down plant matter such as roots, resulting in the accumulation of organic matter. Plants also cause the soil to develop and water retention to increase, therefore providing a habitat where more plants can grow. Vegetation above the high tide line is common on cuspate forelands that are stable and composed of shingle.

Biological Habitat

Cuspate forelands provide a habitat for various flora and fauna. If a foreland is relatively stable and experiences low wave impact, it may be possible for vegetation to grow. In the United Kingdom, 11 taxa of invertebrates are found on shingle habitats. Shingle beaches also provide a habitat for birds to breed, nest, and rest en route while migrating.

Impacts and Management

There are different management issues with regard to cuspate forelands depending on their formation. If a cuspate foreland has formed from deposition, it may be vulnerable if human interference alters the transport of sediments from the shoreline. However if the cuspate foreland is a relic of a past feature that has eroded, human interference with longshore sediment movement will not have a significant impact on the cuspate foreland. For a cuspate foreland to be maintained, the input of sediment must be greater than output of sediment. Activities such as coastal development or engineering must be regulated for sediment to continue moving towards the foreland where it can be deposited. Development along cuspate forelands is risky due to erosion and the vulnerability to storms and sea level rise. As sea levels rise, cuspate forelands are likely to be at risk as they could move inland.

At Point Pelee, approximately 1,900 hectares of former agricultural land on the cuspate foreland is now under water as a result of wind erosion and compaction of organic soils on the foreland. This foreland is particularly vulnerable to erosion when high lake levels are combined with spring and autumn cyclonic activity. Erosion can also occur as spring storms cause ice to scour the lake bottom at the edge of the foreland.

Because there is uncertainty about its formation, there is uncertainty with regard to management, although Parks Canada realises the importance of including Point Pelee National Park in management plans.

When there is an aquifer present under a cuspate foreland, regulation of water removal is required. At Dungeness, water restrictions have been put in place to maintain the aquifer level.

The management of coastlines needs to take into account the natural processes that occur on cuspate forelands since many provide a habitat for birds. Alternative ways of managing coastal erosion is needed, such as the use of 'soft' defences instead of high impact defences such as sea walls.

Some cuspate forelands naturally do not contain any vegetation due to a high level of disturbance from physical factors such as wave action. However with the frequency of storms arising from climate change, the effect on forelands and their associated vegetation needs to be effectively managed.

Mudflat

Mudflats or mud flats, also known as tidal flats, are coastal wetlands that form when mud is deposited by tides or rivers. They are found in sheltered areas such as bays, bayous, lagoons, and estuaries. Mudflats may be viewed geologically as exposed layers of bay mud, resulting from deposition of estuarine silts, clays and marine animal detritus. Most of the sediment within a mudflat is within the intertidal zone, and thus the flat is submerged and exposed approximately twice daily.

In the past tidal flats were considered unhealthy, economically unimportant areas and were often dredged and developed into agricultural land. Several especially shallow mudflat areas, such as the Wadden Sea, are now popular among those practising the sport of mudflat hiking.

On the Baltic Sea coast of Germany in places, mudflats are exposed not by tidal action, but by wind-action driving water away from the shallows into the sea. These wind-affected mudflats are called *windwatts* in German.

Ecology

Tidal flats, along with intertidal salt marshes and mangrove forests, are important ecosystems. They usually support a large population of wildlife, and are a key habitat that allows tens of millions of migratory shorebirds to migrate from breeding sites in the northern hemisphere to non-breeding areas in the southern hemisphere. They are often of vital importance to migratory birds, as well as certain species of crabs, mollusks and fish. In the United Kingdom mudflats have been classified as a Biodiversity Action Plan priority habitat.

The maintenance of mudflats is important in preventing coastal erosion. However, mudflats worldwide are under threat from predicted sea level rises, land claims for development, dredging due to shipping purposes, and chemical pollution.

Geology

Mudflat sediment deposits are focused into the intertidal zone which is composed of a barren zone, marsh and salt pan. Within these areas are various ratios of sand and mud that make up the sedimentary layers. The associated growth of coastal sediment deposits can be attributed to rates of subsidence along with rates of deposition (example: silt transported via river) and changes in sea level.

Barren zones extend from the lowest portion of the intertidal zone to the marsh areas. Beginning in close proximity to the tidal bars, sand dominated layers are prominent and become increasingly muddy throughout the tidal channels. Common bedding types include laminated sand, ripple bedding, and bay mud. Bioturbation also has a strong presence in barren zones.

Marshes contain an abundance of herbaceous plants while the sediment layers consist of thin sand and mud layers. Mudcracks are a common as well as wavy bedding planes. Marshes are also the origins of coal/peat layers because of the abundant decaying plant life.

Salt pans can be distinguished in that they contain thinly laminated layers of clayey silt. The main source of the silt comes from rivers. Dried up mud along with wind erosion forms silt dunes. When flooding, rain or tides come in, the dried sediment is then re-distributed.

Raised Beach

A raised beach, marine terrace, or perched coastline is an emergent coastal landform. Raised beaches and marine terraces are beaches or wave-cut platforms raised above the shore line by a relative fall in the sea level. Around the world, a combination of tectonic coastal uplift and Quaternary sea-level fluctuations has resulted in the formation of marine terrace sequences, most of which were formed during separate interglacial highstands that can be correlated to Marine Oxygen Isotopic Stages (MIS) (for example, Johnson and Libbey (1997).

A marine terrace commonly retains a shoreline angle or inner edge, the slope inflection between the marine abrasion platform and the associated paleo sea-cliff. The shoreline angle represents the maximum shoreline of a transgression and therefore a paleo sea level.

Origin

It is now widely thought that marine terraces are formed during the separated highstands of interglacial stages correlated to Marine Oxygen Isotopic Stages (MIS) (James et al., 1971; Chappell, 1974; Bull, 1985; Ota, 1986; Muhs et al., 1990).

Tectonical and or Eustatical use of Marine Terrace Sequence

The total displacement of the shoreline relative to the age of the associated interglacial stage allows calculation of a mean uplift rate or the calculation of eustatical level at a particular time if the uplift is known.

In order to estimate vertical uplift, the eustatic position of the considered paleo sea levels relative to the present one must be known as precisely as possible. Our chronology relies principally on relative dating based on geomorphologic criteria but in all cases we associated the shoreline angle of the marine terraces with numerical ages. The best-represented terrace worldwide is the one correlated to the last interglacial maximum (MISS 5e) (Hearty and Kindler, 1995; Johnson and Libbey, 1997, Pedoja et al., 2006 a, b, c).

Age of MISS 5e is arbitrarily fixed to range from 130 to 116 ka (Kukla et al., 2002) but is demonstrated to range from 134 to 113 ka in Hawaii and Barbados (Muhs et al., 2002) with a peak from 128 to 116 ka on tectonically stable coastlines (Muhs, 2002). Older marine terraces well represented in worldwide sequences are those related to MIS 9 (~303-339 ka) and 11 (~362-423 ka) (Imbrie et al., 1984). Compilations show that sea level was 3 ± 3 metres higher during MISS 5e, MIS 9 and 11 than during the present one and –1 ± 1 m to the present one during MIS 7 (Hearty and Kindler, 1995, Zazo, 1999). Consequently MIS 7 (~180-240 ka; Imbrie et al., 1984) marine terraces are less pronounced and sometimes absent (Zazo, 1999). When the elevations of these terraces are higher than the uncertainties in paleo-eustatic sea level mentioned for the Holocene and Late Pleistocene, these uncertainties have no effect on overall interpretation.

Sequence can also occurs where the accumulation of ice sheets have depressed the land so that when the ice sheets melts the land readjusts with time thus raising the height of the beaches (glacio-isostatic rebound)and in places where co-seismic uplift occur. In the latter case, the terrace are not correlated with sea level highstand even if co-seismic terrace are known only for the Holocene.

Other Coastal Quaternary Morphologies Registering Uplift

Uplift can also be registered through tidal notch sequences. Notches are often portrayed as lying at sea level; however notch types actually form a continuum from wave notches formed in quiet conditions at sea level to surf notches formed in more turbulent conditions and as much as 2 m (6.6 ft) above sea level (Pirazzoli et al., 1996 in Rust and Kershaw, 2000). As stated above, there was at least one higher sea level during the Holocene, so that some notches may not contain a tectonic component in their formation.

World Wide Occurrence

Raised beaches are found in a wide variety of coast and geodynamical background such as subduction on the pacific coast of South America (Pedoja et al., 2006), of North America, passive margin of the Atlantic coast of South America (Rostami et al., 2000), collision context on the Pacific coast of Kamchatka (Pedoja et al., 2006), Papua New Guinea, New Zealand, Japan (Ota and Yamaguchi, 2004), passive margin of the South China sea coast (Pedoja et al., in press), on west-facing Atlantic coasts, such as Donegal Bay, County Cork and County Kerry in Ireland; Bude, Widemouth Bay, Crackington Haven, Tintagel, Perranporth and St Ives in Cornwall, the Vale of Glamorgan, Gower Peninsula, Pembrokeshire and Cardigan Bay in Wales, the Isle of Jura and Isle of Arran in Scotland, Finistère in Brittany and Galicia in Northern Spain and at Squally Point in the Cape Chignecto Provincial Park, Nova Scotia.

Ria

A ria is a coastal inlet formed by the partial submergence of an unglaciated river valley. It is a drowned river valley that remains open to the sea. Typically, rias have a dendritic, treelike outline although they can be straight and without significant branches. This pattern is inherited from the dendritic drainage pattern of the flooded river valley. The drowning of river valleys along a stretch of coast and formation of rias results in an extremely irregular and indented coastline. Often, there are islands, which are summits of partly submerged, pre-existing hill peaks. A ria coast is a coastline having several parallel rias separated by prominent ridges, extending a distance inland.

The sea level change that caused the submergence of a river valley may be either eustatic (where global sea levels rise), or isostatic (where the local land sinks). The result is often a very large estuary at the mouth of a relatively insignificant river (or else sediments would quickly fill the ria). The Kingsbridge Estuary in Devon, England, is an extreme example of a ria forming an estuary disproportionate to the size of its river; no significant river flows into it at all, only a number of small streams.

Etymology

The word *ría* comes from the Galician language, as rias are present all along the Galician coast. It is related to the word *río* (river). As originally defined, this term was restricted to drowned river valleys cut parallel to the structure of the country rock that was at right angles to the coastline. However, the definition of ria was later expanded to other flooded river valleys regardless of the structure of the country rock. For a period of time, European geomorphologists regarded rias to include any broad estuarine river mouth, including fjords, which are long, narrow inlets with steep sides or cliffs, created in a valley carved by glacial activity. However, the current and preferred usage of this term by geologist and geomorphologists restricts this term solely to drowned unglaciated river valleys and, thus, excludes fjords from being classified as rias.

Strand Plain

A strand plain or strandplain is a broad belt of sand along a shoreline with a surface exhibiting well-defined parallel or semi-parallel sand ridges separated by shallow swales. A strandplain differs from a barrier island in that it lacks either the lagoons or tidales marsh that separate a barrier island from the shoreline to which the strandplain is directly attached. Also, the tidal channels and inlets, which cut through barrier islands, are absent. Strand plains typically are created by the redistribution by waves and longshore currents of coarse sediment on either side of a river mouth. Thus, they are part of one type of wave-dominated delta.

Examples of strand plains:

- Western Louisiana
- Eastern Texas
- West coast of Namibia
- South-east and south-west coasts of Australia, and in the Gulf of Carpentaria

Spit (Landform)

A spit or sandspit is a deposition landform found off coasts. At one end, spits connect to a head, and extend into the nose. A spit is a type of bar or beach that develops where a re-entrant occurs, such as at cove's headlands, by the process of longshore drift. Longshore drift (also called littoral drift) occurs due to waves meeting the beach at an oblique angle, and backwashing perpendicular to the shore, moving sediment down the beach in a zigzag pattern. Longshore drifting is complemented by longshore currents, which transport sediment through the water alongside the beach. These currents are set in motion by the same oblique angle of entering waves that causes littoral drift and transport sediment in a similar process.

Hydrology and Geology

Where the direction of the shore inland *re-enters*, or changes direction, such as at a headland, the longshore current spreads out or dissipates. No longer able to carry the full load, much of the sediment is dropped, called deposition. This submerged bar of sediment allows longshore drift or littoral drift, to continue to transport in the direction the waves are breaking, forming an above-water spit. Without the complementary process of littoral drift, the bar would not build above the surface of the waves becoming a spit and would instead be leveled off underwater.

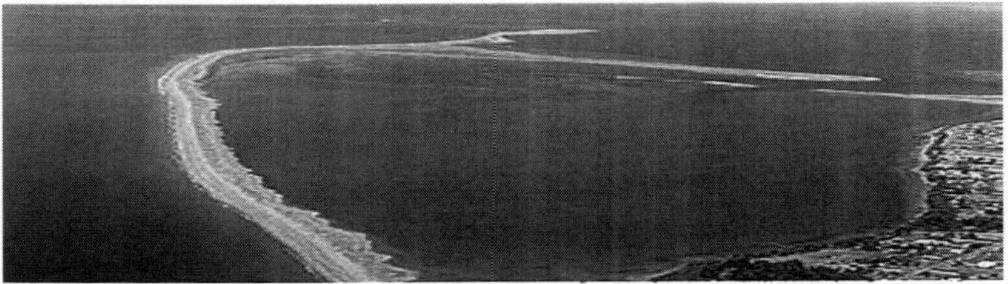

Figure: *Dungeness Spit in the Strait of Juan de Fuca, on the U.S. Pacific coast.*

Spits occur when longshore drift reaches a section of headland where the turn is greater than 30 degrees. They will continue out into the sea until water pressure (such as from a river) becomes too much to allow the sand to deposit. The spit may then be grown upon and become stable and often fertile. A spit may be considered a specialized form of a *shoal.* As spits grow, the water behind them is sheltered from wind and waves, and a salt marsh is likely to develop.

Wave refraction can occur at the end of a spit, carrying sediment around the end to form a hook or recurved spit. Wave refraction in multiple directions will cause a complex spit to form. Incoming waves that come in a direction other than obliquely along the spit will halt the growth of the spit, shorten it or eventually destroy it entirely.

The sediments that make up spits come from a variety of sources including rivers and eroding bluffs, and changes there can have a large impact on spits and other coastal landforms. Activities such as logging and farming upstream can increase the sediment load of rivers, which may hurt the intertidal environments around spits by smothering delicate habitat. Roads or bulkheads built along bluffs can drastically reduce the volume of sediment eroded, so that not enough material is being pushed along to maintain a given spit.

If the supply of sediment is interrupted the sand at the neck (landward side) of the spit may be moved towards the head, eventually creating an island. If the supply isn't interrupted, and the spit isn't breached by the sea (or, if across an estuary, the river) the spit may become a bar, with both ends joined to land, and form a lagoon behind the bar. If an island lies offshore near where the coast changes direction, and the spit continues to grow until it connects the island to the mainland, it is then called a tombolo.

The end of a spit attached to land is called the proximal end, and the end jutting out into water is called the distal end.

Spits Around the World

The longest spit in the world is the Arabat Spit in the Sea of Azov. It is approximately 110 km (68 mi) long.

Farewell Spit in New Zealand, at 32 km (20 mi), in the North West corner of South Island is believed to be caused by the strong prevailing winds and currents bringing sand eroded from the Southern Alps of the South Island and depositing these into Golden Bay.

Spits in the UK are caused by prevailing South-Westerly winds, which give the spits their direction. However, when the direction of the wind changes for a short while the spit may change in direction for a short while forming a hook. Many spits have hooked or curved ends. One spit in the UK can be found in Dorset. Chesil Beach is a 29 km (18 mi) long shingle that connects Weymouth to the Isle of Portland. Chesil Beach provides shelter to Weymouth and the Portland village of Chiswell from prevailing winds and waves. You can also find Spurn Point in Yorkshire and the Humber it is approximately 4.8 km (3.0 mi).

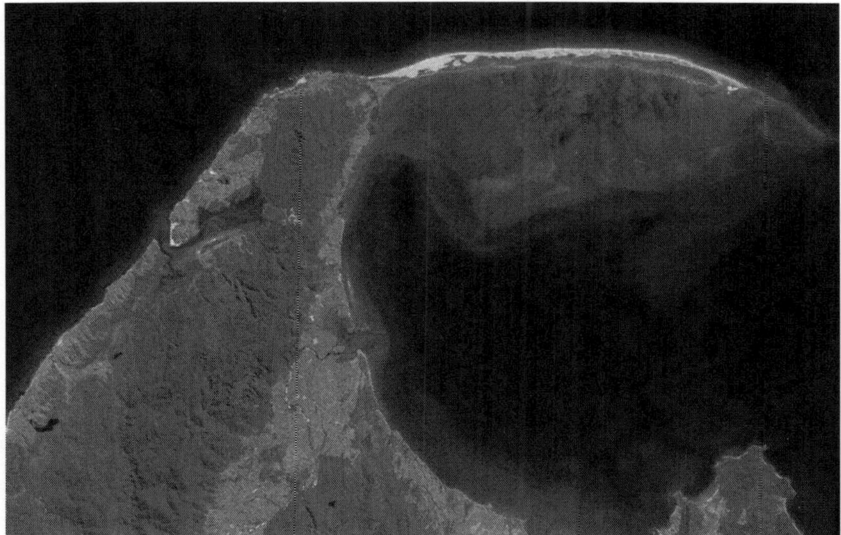

Figure: *Farewell Spit, on New Zealand's South Island*

The Curonian Spit off the coast of Lithuania and Kaliningrad Oblast of Russia separates Curonian Lagoon from the Baltic Sea and is 98 km (61 mi) long. In a similar fashion, the Vistula Spit separates Vistula Lagoon from the Gdańsk Bay off the coast of Poland.

Human Settlement Patterns

Since prehistory humans have chosen certain spit formations as sites for human habitation. In some cases such as Chumash Native American prehistorical settlement, these sites have been chosen for proximity to marine resource exploitation; for example, the Chumash settlement on the Morro Bay sandspit is one such location.

Surge Channel

A surge channel is a narrow inlet on a rocky shoreline. As waves strike the shore, water fills the channel, and drains out again as the waves retreat. The narrow confines of the channel create powerful currents that reverse themselves rapidly as the water level rises and falls.

Surge channels can range from a few inches across to 10 feet or more. They may create tide pools if the conditions are correct, but the rapid water movement almost always creates a dangerous situation for people or animals that are caught in it. The West Coast Trail on the coast of Vancouver Island, B.C., is famous for its large number of surge channels, some of which are impassable even at low tide and must be crossed inland.

Shore

A shore or shoreline is the fringe of land at the edge of a large body of water, such as an ocean, sea, or lake. In physical oceanography, a shore is the wider fringe that is geologically modified by the action of the body of water past and present, while the beach is at the edge of the shore, representing the intertidal zone where there is one. In contrast to a coast, a shore can border any body of water, while the coast must border an ocean; that is, a coast is a type of shore. The word shore is often substituted for coast where an oceanic shore is meant. Shores are influenced by the topography of the surrounding landscape, as well as by water induced erosion, such as waves. The geological composition of rock and soil dictates the type of shore which is created.

Tombolo

A tombolo, from the Italian *tombolo*, derived from the Latin *tumulus*, meaning 'mound,' and sometimes translated as *ayre* (Old Norse *eyrr*, meaning 'gravel beach'), is a deposition landform in which an island is attached to the mainland by a narrow piece of land such as a spit or bar. Once attached, the island is then known as a tied island. Several islands tied together by bars which rise above the water level are called a tombolo cluster. Two or more tombolos may form an enclosure (called a lagoon) that can eventually fill with sediment. Tombolos may be considered a type of isthmus.

Formation

Wave Refraction: True tombolos are formed by wave refraction. As waves near an island, they are slowed by the shallow water surrounding it. These waves then refract or bend around the island to the opposite side as they approach. The wave pattern created by this water movement causes a convergence of longshore drifting on the opposite side of the island. The beach sediments that are moving by lateral transport on the lee side of the island will accumulate there, conforming to the shape of the wave pattern. In other words, the waves sweep sediment together from both sides. Eventually, when enough sediment has built up, the beach shoreline, known as a spit, will connect with an island and form a tombolo.

Lateral Longshore Drift

In the case of Chesil Beach or Spurn Head, the flow of material is along the coast in a movement which is not determined by the now tied island, such as Portland, which it has reached. In this and similar cases, whilst the strip of beach material connected to the island may be technically called a tombolo because it links the island to the land, it is better thought of in terms of its formation- as a spit or otherwise.

Morphology and Sediment Distribution

Tombolos are more prone to natural fluctuations of profile and area as a result of tidal and weather events than a normal beach is. Because of this susceptibility to weathering, tombolos are sometimes made more sturdy through the construction of roads or parking lots. The sediments that make up a tombolo are coarser towards the bottom and finer towards the surface.

It is easy to see this pattern when the waves are destructive and wash away finer grained material at the top, revealing coarser sands and cobbles as the base. Sea level rise may also contribute to accretion, as material is pushed up with rising sea levels. This is the case with Chesil Beach (which connects the Isle of Portland to Dorset in England), notable because the shingle ridge is parallel rather than perpendicular to the coast.

Tombolos demonstrate the sensitivity of shorelines. A small piece of land, such as an island, can change the way that waves move, leading to different deposition of sediments.

Salt Marsh

A salt marsh or saltmarsh, also known as a coastal salt marsh or a tidal marsh, is a coastal ecosystem in the upper coastal intertidal zone between land and open salt water or brackish water that is regularly flooded by the tides. It is dominated by dense stands of salt-tolerant plants such as herbs, grasses, or low shrubs. These plants are terrestrial in origin and are essential to the stability of the salt marsh in trapping and binding sediments. Salt marshes play a large role in the aquatic food web and the delivery of nutrients to coastal waters. They also support terrestrial animals and provide coastal protection.

Basic Information

Figure: *An estuarine salt marsh along the Heathcote River, Christchurch, New Zealand*

Salt marshes occur on low-energy shorelines in temperate and high-latitudes which can be stable or emerging, or submerging if the sedimentation rate exceeds the subsidence rate. Commonly these shorelines consist of mud or sand flats (known also as tidal flats or abbreviated to mudflats) which are nourished with sediment from inflowing rivers and streams. These typically include sheltered environments such as embankments, estuaries and the leeward side of barrier islands and spits. In the tropics and sub-tropics they are replaced by mangroves; an area that differs from a salt marsh in that instead of herbaceous plants, they are dominated by salt-tolerant trees.

Most salt marshes have a low topography with low elevations but a vast wide area, making them hugely popular for human populations. Salt marshes are located among different landforms based on their physical and geomorphological settings. Such marsh landforms include deltaic marshes, estuarine, back-barrier, open coast, embayments and drowned-valley marshes. Deltaic marshes are associated with large rivers where many occur in Southern Europe such as the Camargue, France in the Rhone delta or the Ebro delta in Spain. They are also extensive within the rivers of the Mississippi Delta in the United States. In New Zealand, most salt marshes occur at the head of estuaries in areas where there is little wave action and high sedimentation. Such marshes are located in Awhitu Regional Park in Auckland, the Manawatu Estuary, and the Avon-Heathcote Estuary in Christchurch. Back-barrier marshes are sensitive to the reshaping of barriers in the landward side of which they have been formed. They are common along much of the eastern coast of the United States and the Frisian Islands. Large, shallow coastal embayments can hold salt marshes with examples including Morecambe Bay and Portsmouth in Britain and the Bay of Fundy in North America.

Salt marshes are sometimes included in lagoons, and the difference is not very marked; the Venetian Lagoon in Italy, for example, is made up of these sorts of animals and or living organisms belonging to this ecosystem. They have a big impact on the biodiversity of the area.

Formation

Tidal flats gain elevation relative to sea level by sediment accretion, and subsequently the rate and duration of tidal flooding decreases so that vegetation can colonize on the exposed surface. The arrival of propagules of pioneer species such as seeds or rhizome portions are combined with the development of suitable conditions for their germination and establishment in the process of colonisation. When rivers and streams arrive at the low gradient of the tidal flats, the discharge rate reduces and suspended sediment settles onto the tidal flat surface, helped by the backwater effect of the rising tide. Mats of filamentous blue-green algae can fix silt and clay sized sediment particles to their sticky sheaths on contact which can also increase the erosion resistance of the sediments.

This assists the process of sediment accretion to allow colonising species (e.g., Salicornia spp.) to grow. These species retain sediment washed in from the rising tide around their stems and leaves and form low muddy mounds which eventually coalesce to form depositional terraces, whose upward growth is aided by a sub-surface root network which binds the sediment. Once vegetation is established on depositional terraces further sediment trapping and accretion can allow rapid upward growth of the marsh surface such that there is an associated rapid decrease in the depth and duration of tidal flooding. As a result, competitive species that prefer higher elevations relative to sea level can inhabit the area and often a succession of plant communities develops.

Tidal Flooding and Vegetation Zonation

Figure: *An Atlantic coastal salt marsh in Connecticut.*

Coastal salt marshes can be distinguished from terrestrial habitats by the daily tidal flow that occurs and continuously floods the area. It is an important process in delivering sediments, nutrients and plant water supply to the marsh. At higher elevations in the upper marsh zone, there is much less tidal inflow, resulting in lower salinity levels. Soil salinity in the lower marsh zone is fairly constant due to everyday annual tidal flow.

However, in the upper marsh, variability in salinity is shown as a result of less frequent flooding and climate variations. Rainfall can reduce salinity and evapotranspiration can increase levels during dry periods. As a result, there are microhabitats populated by different species of flora and fauna dependant on their physiological abilities. The flora of a salt marsh is differentiated into levels according to the plants' individual tolerance of salinity and water table levels.

Vegetation found at the water must be able to survive high salt concentrations, periodical submersion, and a certain amount of water movement, while plants further inland in the marsh can sometimes experience dry, low-nutrient conditions. It has been found that the upper marsh zones limit species through competition and the lack of habitat protection, while lower marsh zones are determined through the ability of plants to tolerate physiological stresses such as salinity, water submergence and low oxygen levels.

The New England salt marsh is subject to strong tidal influences and shows distinct patterns of zonation. In low marsh areas with high tidal flooding, a monoculture of the smooth cordgrass, *Spartina alterniflora* dominate, then heading landwards, zones of the salt hay, *Spartina patens*, black rush, *Juncus gerardii* and the shrub *Iva frutescens* are seen respectively. These species all have different tolerances that make the different zones along the marsh best suited for each individual.

Figure: *High marsh in the Marine Park Salt Marsh Nature Centre in Brooklyn, New York*

Plant species diversity is relatively low, since the flora must be tolerant of salt, complete or partial submersion, and anoxic mud substrate. The most common salt marsh plants are glassworts (*Salicornia* spp.) and the cordgrass (*Spartina* spp.), which have worldwide distribution. They are often the first plants to take hold in a mudflat and begin its ecological succession into a salt marsh. Their shoots lift the main flow of the tide above the mud surface while their roots spread into the substrate and stabilize the sticky mud and carry oxygen into it so that other plants can establish themselves as well. Plants such as sea lavenders (*Limonium* spp.), plantains (*Plantago* spp.), and varied sedges and rushes grow once the mud has been vegetated by the pioneer species.

Salt marshes are quite photosynthetically active and are extremely productive habitats. They serve as depositories for a large amount of organic matter and are full of decomposition, which feeds a broad food chain of organisms from bacteria to mammals. Many of the halophytic plants such as cordgrass are not grazed at all by higher animals but die off and decompose to become food for micro-organisms, which in turn become food for fish and birds.

Sediment Trapping, Accretion, and the Role of Tidal Creeks

The factors and processes that influence the rate and spatial distribution of sediment accretion within the salt marsh are numerous. Sediment deposition can occur when marsh species provide a surface for the sediment to adhere to, followed by deposition onto the marsh surface when the sediment flakes off at low tide. The amount of sediment adhering to salt marsh species is dependent on the type of marsh species, the proximity of the species to the sediment supply, the amount of plant biomass, and the elevation of the species. For example, in a study of the Eastern Chongming Island and Jiuduansha Island tidal marshes at the mouth of the Yangtze River, China, the amount of sediment adhering to the species *Spartina alterniflora*, *Phragmites australis*, and *Scirpus mariqueter* decreased with distance from the highest levels of suspended sediment concentrations (found at the marsh edge bordering tidal creeks or the mudflats); decreased with those species at the highest elevations, which experienced the lowest frequency and depth of tidal inundations; and increased with increasing plant biomass. *Spartina alterniflora*, which had the most sediment adhering to it, may contribute >10% of the total marsh surface sediment accretion by this process.

Salt marsh species also facilitate sediment accretion by decreasing current velocities and encouraging sediment to settle out of suspension. Current velocities can be reduced as the stems of tall marsh species induce hydraulic drag, with the effect of minimising re-suspension of sediment and encouraging deposition. Measured concentrations of suspended sediment in the water column have been shown to decrease from the open water or tidal creeks adjacent to the marsh edge, to the marsh interior, probably as a result of direct settling to the marsh surface by the influence of the marsh canopy.

Inundation and sediment deposition on the marsh surface is also assisted by tidal creeks which are a common feature of salt marshes. Their typically dendritic and meandering forms provide avenues for the tide to rise and flood the marsh surface, as well as to drain water, and they may facilitate higher amounts of sediment deposition than salt marsh bordering open ocean. Salt marshes do not however require tidal creeks to facilitate sediment flux over their surface although salt marshes with this morphology seem to be rarely studied.

The elevation of marsh species is important; those species at lower elevations experience longer and more frequent tidal floods and therefore have the opportunity for more sediment deposition to occur.

Species at higher elevations can benefit from a greater chance of inundation at the highest tides when increased water depths and marsh surface flows can penetrate into the marsh interior.

Human Impacts

The coast is a highly attractive natural feature to humans through its beauty, resources, and accessibility. As of 2002, over half of the world's population was estimated to being living within 60 km of the coastal shoreline, making coastlines highly vulnerable to human impacts from daily activities that put pressure on these surrounding natural environments. In the past, salt marshes were perceived as coastal 'wastelands,' causing considerable loss and change of these ecosystems through land reclamation for agriculture, urban development, salt production and recreation. The indirect effects of human activities such as nitrogen loading also play a major role in the salt marsh area.

Land Reclamation

Reclamation of land for agriculture by converting marshland to upland was historically a common practice. Dikes were often built to allow for this shift in land change and to provide flood protection further inland. In recent times intertidal flats have also been reclaimed. For centuries, livestock such as sheep and cattle grazed on the highly fertile salt marsh land. Land reclamation for agriculture has resulted in many changes such as shifts in vegetation structure, sedimentation, salinity, water flow, biodiversity loss and high nutrient inputs. There have been many attempts made to eradicate these problems for example, in New Zealand, the cordgrass *Spartina anglica* was introduced from England into the Manawatu River mouth in 1913 to try and reclaim the estuary land for farming. A shift in structure from bare tidal flat to pastureland resulted from increased sedimentation and the cordgrass extended out into other estuaries around New Zealand. Native plants and animals struggled to survive as non-natives out competed them. Efforts are now being made to remove these cordgrass species, as the damages are slowly being recognised.

In the Blyth estuary, Suffolk, southeast England, the mid-estuary reclamations (Angel and Bulcamp marshes) that were abandoned in the 1940s have been replaced by tidal flats with compacted soils from agricultural use overlain with a thin veneer of mud. Little vegetation colonisation has occurred in the last 60–75 years and has been attributed to a combination of surface elevations too low for pioneer species to develop, and poor drainage from the compacted agricultural soils acting as an aquaclude. Terrestrial soils of this nature need to adjust from fresh to saline interstitial water by a change in the chemistry and the structure of the soil, accompanied with fresh deposition of estuarine sediment, before salt marsh vegetation can establish. The vegetation structure, species richness, and plant community composition of salt marshes naturally regenerated on reclaimed agricultural land can be compared to adjacent reference salt marshes to assess the success of marsh regeneration.

Upstream Agriculture

Cultivation of land upstream from the salt marsh can introduce increased silt inputs and raise the rate of primary sediment accretion on the tidal flats, so that pioneer species can spread further onto the flats and grow rapidly upwards out of the level of tidal inundation. As a result, marsh surfaces in this regime may have an extensive cliff at their seaward edge. At the Plum Island estuary, Massachusetts (U.S.A), stratigraphic cores revealed that during the 18th and 19th century the marsh prograded over subtidal and mudflat environments to increase in area from 6 km^2 to 9 km^2 after European settlers deforested the land uptream and increased the rate of sediment supply.

Urban Development and Nitrogen Loading

The conversion of marshland to upland for agriculture has in the past century been overshadowed by conversion for urban development. Coastal cities worldwide have encroached onto former salt marshes and in the U.S. the growth of cities looked to salt marshes for waste disposal sites. Estuarine pollution from organic, inorganic, and toxic substances from urban development or industrialisation is a worldwide problem and the sediment in salt marshes may entrain this pollution with toxic effects on floral and faunal species. Urban development of salt marshes has slowed since about 1970 owing to growing awareness by environmental groups that they provide beneficial ecosystem services. They are highly productive ecosystems, and when net productivity is measured in g m^{-2} yr^{-1} they are equalled only by tropical rainforests. Additionally, they can help reduce wave erosion on sea walls designed to protect low-lying areas of land from wave erosion. De-naturalisation of the landward boundaries of salt marshes from urban or industrial enchroachment can have negative effects. In the Avon-Heathcote estuary/Ihutai, New Zealand, species abundance and the physical properties of the surrounding margins were strongly linked, and the majority of salt marsh was found to be living along areas with natural margins in the Avon and Heathcote river outlets; conversely, artificial margins contained little marsh vegetation and restricted landward retreat. The remaining marshes surrounding these urban areas are also under immense pressure from the human population as human-induced nitrogen enrichment enters these habitats. Nitrogen loading through human-use indirectly affects salt marshes causing shifts in vegetation structure and the invasion of non-native species. Human impacts such as sewage, urban run-off, agricultural and industrial wastes are running into the marshes from nearby sources.

Salt marshes are nitrogen limited and with an increasing level of nutrients entering the system from anthropogenic effects, the plant species associated with salt marshes are being restructured through change in competition. For example, the New England salt marsh is experiencing a shift in vegetation structure where *S. alterniflora* is spreading from the lower marsh where it predominately resides up into the upper marsh zone. Additionally, in the same marshes, the reed *Phragmites australis* has been invading the area expanding to lower marshes and becoming a dominant species. *P. australis* is an aggressive halophyte that can invade disturbed areas in large numbers outcompeting native plants. This loss in biodiversity is not only seen in flora assemblages but also in many animals such as insects and birds as their habitat and food resources are altered.

Mosquito Control

Earlier in the 20th century, it was believed that draining salt marshes would help reduce mosquito populations. In many locations, particularly in the northeastern United States, residents and local and state agencies dug straight-lined ditches deep into the marsh flats. The end result, however, was a depletion of killifish habitat. The killifish is a mosquito predator, so the loss of habitat actually led to higher mosquito populations, and adversely affected wading birds that preyed on the killifish. These ditches can still be seen, despite some efforts to refill the ditches.

Crab Herbivory and Bioturbation

Increased nitrogen uptake by marsh species into their leaves can prompt greater rates of length-specific leaf growth, and increase the herbivory rates of crabs. The burrowing crab *Neohelice granulata* frequents SW Atlantic salt marshes where high density populations can be found among populations of the marsh species *Spartina densiflora* and *Sarcocornia perennis*. In Mar Chiquita lagoon, north of Mar del Plata, Argentina, *Neohelice granulata* herbivory increased as a likely response to the increased nutrient value of the leaves of fertilized *Spartina densiflora* plots, compared to non-fertilized plots. Regardless of whether the plots were fertilized or not, grazing by *Neohelice granulata* also reduced the length specific leaf growth rates of the leaves in summer, while increasing their length-specific senescence rates. This may have been assisted by the increased fungal effectiveness on the wounds left by the crabs. The salt marshes of Cape Cod, Massachusetts (U.S.A), are experiencing creek bank die-offs of *Spartina* spp. (cordgrass) that has been attributed to herbivory by the crab *Sesarma reticulatum*.

At 12 surveyed Cape Cod salt marsh sites, 10% - 90% of creek banks experienced die-off of cordgrass in association with a highly denuded substrate and high density of crab burrows. Populations of *Sesarma reticulatum* are increasing, possibly as a result of the degradation of the coastal food web in the region. The bare areas left by the intense grazing of cordgrass by *Sesarma reticulatum* at Cape Cod are suitable for occupation by another burrowing crab, *Uca pugnax*, which are not known to consume live macrophytes. The intense bioturbation of salt marsh sediments from this crab's burrowing activity has been shown to dramatically reduce the success of *Spartina alterniflora* and *Suaeda maritima* seed germination and established seedling survival, either by burial or exposure of seeds, or uprooting or burial of established seedlings.

However, bioturbation by crabs may also have a positive effect. In New Zealand, the tunnelling mud crab *Helice crassa* has been given the stately name of an 'ecosystem engineer' for its ability to construct new habitats and alter the access of nutrients to other species. Their burrows provide an avenue for the transport of dissolved oxygen in the burrow water through the oxic sediment of the burrow walls and into the surrounding anoxic sediment, which creates the perfect habitat for special nitrogen cycling bacteria. These nitrate reducing (denitrifying) bacteria quickly consume the dissolved oxygen entering into the burrow walls to create the oxic mud layer that is thinner than that at the mud surface. This allows a more direct diffusion path for the export of nitrogen (in the form of gaseous nitrogen (N_2)) into the flushing tidal water.

Restoration and Management

The perception of bay salt marshes as a coastal 'wasteland' has since changed, acknowledging that they are one of the most biologically productive habitats on earth, rivalling tropical rainforests. Salt marshes are ecologically important providing habitats for native migratory fish and acting as sheltered feeding and nursery grounds. They are now protected by legislation in many countries to look after these ecologically important habitats. In the United States and Europe, they are now accorded to a high level of protection by the Clean Water Act and the Habitats Directive respectively. With the impacts of this habitat and its importance now realised, a growing interest in restoring salt marshes, through managed retreat or the reclamation of land has been established. However, many Asian countries such as China are still to recognise the value of marshlands. With their ever-growing populations and intense development along the coast, the value of salt marshes tends to be ignored and the land continues to be reclaimed.

Bakker et al. (1997) suggests two options available for restoring salt marshes. The first is to abandon all human interference and leave the salt marsh to complete its natural development. These types of restoration projects are often unsuccessful as vegetation tends to struggle to revert to its original structure and the natural tidal cycles are shifted due to land changes. The second option suggested by Bakker et al. (1997) is to restore the destroyed habitat into its natural state either at the original site or as a replacement at a different site. Under natural conditions, recovery can take 2–10 years or even longer depending on the nature and degree of the disturbance and the relative maturity of the marsh involved. Marshes in their pioneer stages of development will recover more rapidly than mature marshes as they are often first to colonize the land. It is important to note, that restoration can often be sped up through the replanting of native vegetation.

This last approach is often the most practiced and generally more successful than allowing the area to naturally recover on its own. The salt marshes in the state of Connecticut in the United States have long been an area lost to fill and dredging. As of 1969, the Tidal Wetland Act was introduced that seized this practice, but despite the introduction of the act, the system was still degrading due to alterations in tidal flow. One area in Connecticut is the marshes on Barn Island. These marshes were diked then impounded with salt and brackish marsh during 1946-1966. As a result the marsh shifted to a freshwater state and became dominated by the invasive species *P. australis*, *Typha angustifolia* and *T. latifolia* that have little ecological connection to the area. By 1980, a restoration programme was put in place that has now been running for over 20 years. This programme has aimed to reconnect the marshes by returning tidal flow along with the ecological functions and characteristics of the marshes back to their original state. In the case of Barn Island, declines in the invasive species have initiated, re-establishing the tidal-marsh vegetation along with animal species such as fish and insects. This example highlights that considerable time and effort is needed to effectively restore salt marsh systems. Times in marsh recovery can depend on the development stage of the marsh; type and extent of the disturbance; geographical location; and the environmental and physiological stress factors to the marsh-associated flora and fauna.

Although much effort has gone into restoring salt marshes worldwide, further research is needed. There are many setbacks and problems associated with marsh restoration that requires careful long-term monitoring. Information on all components of the salt marsh ecosystem should be understood and monitored from sedimentation, nutrient, and tidal influences, to behaviour patterns and tolerances of both flora and fauna species. Once we have a better understanding of these processes and not just locally, but over a global scale, we can then suggest more sound and practical management and restoration efforts that can be used to preserve our valuable marshes and put them back to their original state.

While humans are situated along coastlines, there will always be the possibility of human-induced disturbances despite the number of restoration efforts we plan to implement. Dredging, pipelines for offshore petroleum resources, highway construction, accidental toxic spills or just plain carelessness are examples that will for some time now and into the future be the major influences of salt marsh degradation.

In addition to restoring and managing salt marsh systems based on scientific principles, the opportunity should be taken to educate public audiences of their importance biologically and their purpose as serving as a natural buffer for flood protection. Because salt marshes are often located next to urban areas, they are likely to receive more visitors than remote wetlands. By physically seeing the marsh, people are more likely to take notice and be more aware of the environment around them. An example of public involvement occurred at the Famosa Slough wetland in San Diego, where a "friends" group worked for over a decade in trying to prevent the area from being developed. Eventually, the 5 hectare site was bought by the City and the group worked together to restore the area. The project involved removing invasive species and replanting with natives, along with public talks to other locals, frequent bird walks and clean-up events.

Research Methods

There is a diverse range and combination of methodologies employed to understand the hydrological dynamics in salt marshes and their ability to trap and accrete sediment. Sediment traps are often used to measure rates of marsh surface accretion when short term deployments (e.g. less than one month) are required. These circular traps consist of pre-weighed filters that are anchored to the marsh surface, then dried in a laboratory and re-weighed to determine the total deposited sediment. For longer term studies (e.g. more than one year) researchers may prefer to measure sediment accretion with marker horizon plots. Marker horizons consist of a mineral such as feldspar that is buried at a known depth within wetland substrates to record the increase in overlying substrate over long time periods. In order to gauge the amount of sediment suspended in the water column, manual or automated samples of tidal water can be poured through pre-weighed filters in a laboratory then dried to determine the amount of sediment per volume of water. Another method for estimating suspended sediment concentrations is by measuring the turbidity of the water using optical backscatter probes, which can be calibrated against water samples containing a known suspended sediment concentration to establish a regression relationship between the two. Marsh surface elevations may be measured with a stadia rod and transit, electronic theodolite, Real-Time Kinematic Global Positioning System, laser level or electronic distance metre (total station). Hydrological dynamics include water depth, measured automatically with a pressure transducer, or with a marked wooden stake, and water velocity, often using electromagnetic current metres.

Coastal Management

In some jurisdictions the terms sea defence and coastal protection are used to mean, respectively, defence against flooding and erosion. The term *coastal defence* is the more traditional term, but *coastal management* has become more popular as the field has expanded to include techniques that allow erosion to claim land.

Historical Background

Coastal engineering, as it relates to harbours, starts with the development of ancient civilizations together with the origin of maritime traffic, perhaps before 3500 B.C.

Docks, breakwaters, and other harbour works were built by hand and often in a grand scale. Basic source of modern literature on coastal engineering is the "European Code of Conduct for Coastal Zones" issued by the European Council in 1999. This document was prepared by the Group of Specialists on Coastal Protection and should be used 'as a source of inspiration for national legislation and practice' by decision makers.

The Group of Specialists on Coastal Protection (PE-S-CO), set up in 1995, pursuant to a decision by the Committee of Ministers of the Council of Europe, met for the first time on 6 and 7 June 1996. It noted that a great deal of technical and scientific research had been carried out in the field of coastal protection and that various principles and legal texts had been drawn up. It also noted that all of the work undertaken highlighted the need for integrated management and planning of coastal areas, but that, despite all the efforts already made, the situation of coastal areas continued to deteriorate.

The Group had acknowledged that this was due to difficulties in implementing the concept of "integrated management", and that it was becoming necessary to provide instruments which would make it easier to apply the principles of integrated coastal management and planning, which had to be pursued to ensure sustainable management of coastal areas. The Group therefore proposed that the Council of Europe, in close co-operation with the European Union for Coastal Conservation (EUCC) and United Nations Environment Programme (UNEP). The final version of the Code as well of the a MODEL LAW to be used as a guide for modifying local and national legislation, can be free downloaded from the web.

Ancient harbour works are still visible in a few of the harbours that exist today, while others have recently been explored by underwater archaeologists. Most of the grander ancient harbor works have disappeared following the fall of the Roman Empire.

Most ancient coastal efforts were directed to port structures, with the exception of a few places where life depended on coastline protection. Venice and its lagoon is one such case. Protection of the shore in Italy, England and the Netherlands can be traced back at least to the 6th century. The ancients understood such phenomena as the Mediterranean currents and wind patterns and the wind-wave cause-effect link.

The Romans introduced many revolutionary innovations in harbor design. They learned to build walls underwater and managed to construct solid breakwaters to protect fully exposed harbours. In some cases wave reflection may have been used to prevent silting. They also used low, water-surface breakwaters to trip the waves before they reached the main breakwater. They became the first dredgers in the Netherlands to maintain the harbour at Velsen. Silting problems here were solved when the previously sealed solid piers were replaced with new "open"-piled jetties.

Middle Ages

The threat of attack from the sea caused many coastal towns and their harbours to be abandoned. Other harbours were lost due to natural causes such as rapid silting, shoreline advance or retreat, etc. The Venetian Lagoon was one of the few populated coastal areas with continuous prosperity and development where written reports document the evolution of coastal protection works. Engineering and scientific skills remained alive in the east, in Byzantium, where the Eastern Roman Empire survived for six hundred years while Western Rome decayed.

Modern Age

Although great strides were made in the general scientific arena, little improvement was done beyond the Roman approach to harbour construction after the Renaissance. In the early 19th century, the advent of the steam engine, the search for new lands and trade routes, the expansion of the British Empire through her colonies, and other influences, all contributed to the revitalization of sea trade and a renewed interest in port works.

Twentieth Century

Evolution of shore protection and the shift from structures to beach nourishment. Prior to the 1950s, the general practice was to use hard structures to protect against beach erosion or storm damages. These structures were usually coastal armoring such as seawalls and revetments or sand-trapping structures such as groynes. During the 1920s and '30s, private or local community interests protected many areas of the shore using these techniques in a rather ad hoc manner. In certain resort areas, structures had proliferated to such an extent that the protection actually impeded the recreational use of the beaches. Erosion of the sand continued, but the fixed back-beach line remained, resulting in a loss of beach area.

The obtrusiveness and cost of these structures led in the late 1940s and early 1950s, to move toward a new, more dynamic, method.

Projects no longer relied solely on hard coastal defence structures, as techniques were developed which replicated the protective characteristics of natural beach and dune systems. The resultant use of artificial beaches and stabilized dunes as an engineering approach was an economically viable and more environmentally friendly means for dissipating wave energy and protecting coastal developments.

Over the past hundred years the limited knowledge of coastal sediment transport processes at the local authorities level has often resulted in inappropriate measures of coastal erosion mitigation. In many cases, measures may have solved coastal erosion locally but have exacerbated coastal erosion problems at other locations -up to tens of kilometres away- or have generated other environmental problems.

Current Challenges in Coastal Management

The coastal zone is a dynamic equilibrium area of natural change and of increasing human use. They occupy less than 15% of the Earth's land surface; yet accommodate more than 40% of the world population (it is estimated that 3.1 billion people live within 200 kilometres from the sea). With three-quarters of the world population expected to reside in the coastal zone by 2025, human activities originating from this small land area will impose an inordinate amount of pressures on the global system. Coastal zones contain rich resources to produce goods and services and are home to most commercial and industrial activities. In the European Union, almost half of the population now lives within 50 kilometres of the sea and coastal zone resources produce much of the Union's economic wealth.

The fishing, shipping and tourism industries all compete for vital space along Europe's estimated 89 000 kilometres of coastline, and coastal zones contain some of Europe's most fragile and valuable natural habitats. Shore protection consists up to the 50's of interposing a static structure between the sea and the land to prevent erosion and or flooding, and it has a long history. From that period new technical or friendly policies have been developed to preserve the environment when possible.

Is already important where there are extensive low-lying areas that require protection. For instance: Venice, New Orleans, Nagara river in Japan, the Netherlands, Caspian Sea Protection against the sea level rise in the 21st century will be especially important, as sea level rise is currently accelerating. This will be a challenge to coastal management, since seawalls and breakwaters are generally expensive to construct, and the costs to build protection in the face of sea-level rise would be enormous.

Changes on sea level have a direct adaptative response from beaches and coastal systems. When the sea level rises, coastal sediments are in part pushed up by wave and tide energy, so sea-level rise processes have a component of sediment transport landwards. This results in a dynamic model of rise effects with a continuous sediment displacement that is not compatible with static models where coastline change is only based on topographic data.

Planning Approaches

There are five generic strategies for coastal defence:

- inaction leading to eventual abandonment
- Managed retreat or realignment, which plans for retreat and adopts engineering solutions that recognise natural processes of adjustment, and identifies a new line of defence where to construct new defences
- Hold the line, shoreline protection, whereby seawalls are constructed around the coastlines
- Move seawards, this happens by constructing new defences seaward the original ones
- Limited intervention, accommodation, by which adjustments are made to be able to cope with inundation, raising coastal land and buildings vertically

The decision to choose a strategy is site-specific, depending on pattern of relative sea-level change, geomorphological setting, sediment availability and erosion, as well a series of social, economic and political factors.

Alternatively, integrated coastal zone management approaches may be used to prevent development in erosion- or flood-prone areas to begin with. Growth management can be a challenge for coastal local authorities who often struggle to provide the infrastructure required by new residents seeking seachange lifestyles. Sustainable transport investment to reduce the average footprint of coastal visitors is often a good way out of coastal gridlock. Examples include Dongtan and the Gold Coast Oceanway.

The 'Managed Retreat' option, involving no protection, is cheap and expedient. The coast takes care of itself and coastal facilities are abandoned to coastal erosion, with either gradual landward retreat or evacuation and resettlement elsewhere. This is the usual response when land of little value will be lost. The only pollution produced is from the resettlement process. Where endangered property has high value, it is less often applied.

Managed Retreat

Managed retreat is an alternative to constructing or maintaining coastal structures. Managed retreat allows an area to become flooded. This process is usually in low lying estuarine or deltaic areas and floods land that has at some point in the past been reclaimed from the sea. Managed retreat is often a response to a change in sediment budget or to sea level rise. The technique is used when the land adjacent to the sea is low in value. A decision is made to allow the land to erode and flood, creating new shoreline habitats. This process may continue over many years and natural stabilization will occur.

The earliest managed retreat in the UK was an area of 0.8 ha at Northey Island in Essex, that was flooded in 1991. This was followed by Tollesbury and Orplands in Essex, where the sea walls were breached in 1995. In the Ebro delta (Spain) coastal authorities have planned a managed retreat in response to coastal erosion (MMA 2005, Sitges, Meeting on Coastal Engineering; EUROSION project).

Cost – The main cost is generally the purchase of land to be flooded. Compensation for relocation of residents may be needed. Any other human made structure which will be engulfed by the sea may need to be safely dismantled to prevent sea pollution. In some cases, a retaining wall or bund must be constructed inland in order to protect land beyond the area to be flooded, although such structures can generally be lower than would be needed on the existing coast. Monitoring of the evolution of the flooded area is another cost. Costs may be lowest if existing defences are left to fail naturally, but often the realignment project will be more actively managed, for example by creating an artificial breach in existing defences to allow the sea in at a particular place in a controlled fashion, or by pre-forming drainage channels for created salt-marsh.

Hold the Line

Human strategies on the coast have been heavily based on a static engineered response, whereas the coast is in, or strives towards, a dynamic equilibrium (Schembri, 2009). Solid coastal structures are built and persist because they protect expensive properties or infrastructures, but they often relocate the problem downdrift or to another part of the coast. Soft options like beach nourishment, while also being temporary and needing regular replenishment, appear more acceptable, and go some way to restore the natural dynamism of the shoreline. However in many cases there is a legacy of decisions that were made in the past which have given rise to the present threats to coastal infrastructure and which necessitate immediate shore protection. For instance, the seawall and promenade of many coastal cities in Europe represents a highly engineered use of prime seafront space, which might be preferably designated as public open space, parkland and amenities if it were available today.

Such open space might also allow greater flexibility in terms of future land-use change, for instance through managed retreat, in the face of threats of erosion or inundation as a result of sea-level rise. Foredunes areas represent a natural reserve which can be called upon in the face of extreme events; building on these areas leaves little option but to undertake costly protective measures when extreme events (whether amplified by gradual global change or not) threaten. Managed retreat can comprise 'setbacks', rolling easements and other planning tools including building within a particular design life. Maintenance of those structures or soft techniques can arrive at a critical point (economically or environmental) to change adopted strategy.

- Structural or hard engineering techniques, i.e. using permanent concrete and rock constructions to "fix" the coastline and protect the assets locate behind. These techniques—seawalls, groynes, detached breakwaters, and revetments—represent a significant share of protected shoreline in Europe (more than 70%).
- Soft engineering techniques (e.g. sand nourishments), building with natural processes and relying on natural elements such as sands, dunes and vegetation to prevent erosive forces from reaching the backshore. These techniques include beach nourishment and sand dune stabilization.

Move Seaward

The futility of trying to predict future scenarios where there is a large human influence is apparent. Even future climate is to a certain extent a function of what humans choose to make of it, for example by restricting greenhouse gas emissions to control climate change. In some cases - where new areas are needed for new economic or ecological development - a move seaward strategy can be adopted. Examples from erosion include: Koge Bay (Dk) Western Scheldt estuary (Nl), Chatelaillon (F), Ebro delta (E)

There is an obvious downside to this strategy. Coastal erosion is already widespread, and there are many coasts where exceptional high tides or storm surges result in encroachment on the shore, impinging on human activity. If the sea rises, many coasts that are developed with infrastructure along or close to the shoreline will be unable to accommodate erosion. They will experience a so-called "coastal squeeze" whereby the ecological or geomorphological zones that would normally retreat landwards encounter solid structures and are squeezed out. Wetlands, salt marshes, mangroves and adjacent fresh water wetlands are particularly likely to suffer from this squeeze.

An upside to the strategy is that moving seaward (and upward) can create land of high value which can bring the investment required to cope with climate change.

Limited Intervention

Limited intervention is an action taken whereby the management only solves the problem to some extent, usually in areas of low economic significance. Measures taken using limited intervention often encourage the succession of haloseres, including salt marshes and sand dunes. This will normally result in the land behind the halosere being more sufficiently protected, as wave energy will be dissipated by the accumulated sediment and additional vegetation residing in the newly formed habitat. Although the new halosere is not strictly man-made, as many natural processes will contribute to the succession of the halosere, anthropogenic factors are partially responsible for the formation as an initial factor was needed to help start the process of succession. This must not be confused with 'accommodate' which is about property e.g. effective insurance, early warning systems and not about habitat.

Construction Techniques

The following is a catalogue of relevant techniques that could be employed as coastal management techniques. *The costs given are very rough estimates made during 2005, based on UK Pound sterling.*

3

Hard Engineering Methods

Groyne

A groyne (groin in the United States) is a rigid hydraulic structure built from an ocean shore (in coastal engineering) or from a bank (in rivers) that interrupts water flow and limits the movement of sediment. In the ocean, groynes create beaches, or prevent them being washed away by longshore drift. In a river, groynes prevent erosion and ice-jamming, which in turn aids navigation. Ocean groynes run generally perpendicular to the shore, extending from the upper foreshore or beach into the water. All of a groyne may be under water, in which case it is a submerged groyne. The areas between groups of groynes are groyne fields. Groynes are generally made of wood, concrete, or rock piles, and placed in groups. They are often used in tandem with seawalls. Groynes, however, may cause a shoreline to be perceived as unnatural.

In Coastal Engineering

A groyne's length and elevation, and the spacing between groynes is determined according to local wave energy and beach slope. Groynes that are too long or too high tend to accelerate downdrift erosion because they trap too much sediment. Groynes that are too short, too low, or too permeable are ineffective because they trap too little sediment. Flanking may occur if a groyne does not extend far enough landward.

Working

A groyne creates and maintains a wide area of beach or sediment on its updrift side, and reduces erosion on the other. It is a physical barrier to stop sediment transport in the direction of longshore drift (also called longshore transport). This causes a build-up, which is often accompanied by accelerated erosion of the downdrift beach, which receives little or no sand from longshore drift (this is known as terminal groyne syndrome, as it occurs after the terminal groyne in a group of groynes).

Groynes do not add extra material to a beach, but merely retain some of the existing sediment on the updrift side of the groynes. If a groyne is correctly designed, then the amount of material it can hold will be limited, and excess sediment will be free to move on through the system. However, if a groyne is too large it may trap too much sediment, which can cause severe beach erosion on the down-drift side.

In Rivers

River groynes (spur dykes or wing dykes) (American English: "dikes") are often constructed nearly perpendicular to the riverbanks, beginning at a riverbank with a root and ending at the regulation line with a head. They maintain a channel to prevent ice jamming, and more generally improve navigation and control over lateral erosion, that would form from meanders. Groynes have a major impact on the river morphology: they cause autonomous degradation of the river.

They are also used around bridges to prevent bridge scour.

Types

Groynes can be distinguished by how they are constructed, whether they are submerged, their effect on stream flow or by shape.

By Construction Method

Groynes can be permeable, allowing the water to flow through at reduced velocities, or impermeable, blocking and deflecting the current.

- Permeable groynes are large rocks, bamboo or timber
- impermeable groynes (solid groynes or rock armour groynes) are constructed using rock, gravel, gabions.

By whether they are submerged

Groynes can be submerged or not under normal conditions. Usually impermeable groynes are non-submerged, since flow over the top of solid groynes may cause severe erosion along the shanks. Submerged groynes, on the other hand, may be permeable depending on the degree of flow disturbance needed.

By Their Effect on Stream Flow

Groynes can be attracting, deflecting or repelling.

- Attracting groynes point downstream, serving to attract the stream flow towards themselves and not repel the flow towards the opposite bank. They tend to maintain deep current close to the bank.
- Deflecting groynes change the direction of flow without repelling it. They are generally short and used for limited, local protection.
- Repelling groynes point upstream; they force the flow away from themselves. A single groyne may have one section, for example, attracting, and another section deflecting.

By Shape

Groynes can be built with different planview shapes. Examples are straight groynes, T head, L head, hockey stick, inverted hockey stick groynes, straight groynes with pier head, wing, and tail groynes.

Seawall

A seawall (or sea wall) is a form of coastal defence constructed where the sea, and associated coastal processes, impact directly upon the landforms of the coast. The purpose of a seawall is to protect areas of human habitation, conservation and leisure activities from the action of tides and waves. As a seawall is a static feature it will conflict with the dynamic nature of the coast and impede the exchange of sediment between land and sea.

The coast is generally a high-energy, dynamic environment with spatial variations occurring over a wide range of temporal scales. The shoreline is part of the coastal interface which is exposed to a wide range of erosional processes arising from fluvial, aeolian and terrestrial sources, meaning that a combination of denudational processes will work against a seawall. Given the natural forces to which seawalls are constantly subjected, maintenance (and eventually replacement) is an ongoing requirement if they are to provide an effective long term solution.

The many types of seawall in use today reflect both the varying physical forces they are designed to withstand, and location specific aspects, such as: local climate, coastal position, wave regime, and value of landform. Seawalls are classified as a hard engineering shore based structure used to provide protection and to lessen coastal erosion. However, a range of environmental problems and issues may arise from the construction of a seawall, including disrupting sediment movement and transport patterns, which are discussed in more detail below. Combined with a high construction cost, this has led to an increasing use of other soft engineering coastal management options such as beach replenishment.

Seawalls may be constructed from a variety of materials, most commonly: reinforced concrete, boulders, steel, or gabions. Additional seawall construction materials may include: vinyl, wood, aluminium, fibreglass composite, and with large biodegrable sandbags made of jute and coir. In the UK, *sea wall* also refers to an earthen bank used to create a polder, or a dike construction.

Trade-offs

A cost benefit approach is an effective way to determine whether a seawall is appropriate and whether the benefits are worth the expense. Besides controlling erosion, consideration must be given to the effects of hardening a shoreline on natural coastal ecosystems and human property or activities. A seawall is a static feature which can conflict with the dynamic nature of the coast and impede the exchange of sediment between land and sea. The table below summarises some positive and negative effects of seawalls which can be used when comparing their effectiveness with other coastal management options, such as beach nourishment.

Advantages and disadvantages of seawalls according to Short (1999)	
Advantages	***Disadvantages***
• Long term solution in comparison to soft beach nourishment. • Effectively minimizes loss of life in extreme events and damage to property caused by erosion. • Can exist longer in high energy environments in comparison to 'soft' engineering methods. • Can be used for recreation and sightseeing. • Forms a hard and strong coastal defence.	• Very expensive to construct. • Can cause beaches to dissipate rendering them useless for beach goers. • Scars the very landscape that they are trying to save and provides an 'eyesore.' • Reflected energy of waves leading to scour at base. • Can disrupt natural shoreline processes and destroy shoreline habitats such as wetlands and intertidal beaches. • Altered sediment transport processes can disrupt sand movement that can lead to increased erosion down drift from the structure.

Generally seawalls can be a successful way to control coastal erosion, but only if they are constructed well and out of materials which can withstand the force of ongoing wave energy. Some understanding is needed of the coastal processes and morphodynamics specific to the seawall location. Seawalls can be very helpful; they can offer a more long term solution than soft engineering options, additionally providing recreation opportunities and protection from extreme events as well as everyday erosion. Extreme natural events expose weaknesses in the performance of seawalls, and analyses of these can lead to future improvements and reassessment.

Simulations

In 2007 researchers at the University of Salerno published studies showing interactions between maritime breakwaters and waves using CAD and CFD software. In the simulations the filtration motion of the fluid within the interstices, which normally exist in a breakwater, is estimated by integrating the relevant RANS equations coupled with a random number generated turbulence model inside the voids, rather than using the classical equations for porous media. The breakwaters were modelled, both for the actual size construction and for a physical laboratory test, by overlapping three-dimensional elements. The numerical grid was thickened in such a way to have some computational nodes along the flow paths among the breakwater's blocks.

Issues

Sea Level Rise: Sea level rise creates an issue for seawalls worldwide as it raises both the mean normal water level and the height of waves during extreme weather events, which the current seawall heights may be unable to cope with (Allan *et al.* 1999). The International Panel on Climate Change (IPCC) (1997) suggested that sea level rise over the next 50 – 100 years will accelerate with a projected increase in global mean sea level of +18 cm by 2050 AD. This data is reinforced by Hannah (1990) who calculated similar statistics including a rise of between +16-19.3 cm throughout 1900–1988. This problem could be overcome by further modelling and determining the extension of height and reinforcement of current seawalls which needs to occur for safety to be ensured in both situations.

Extreme events

Extreme events also pose a problem as it is not easy for people to predict or imagine the strength of hurricane or storm induced waves compared to normal, expected wave patterns. An extreme event can dissipate hundreds of times more energy than everyday waves, and calculating structures which will stand the force of coastal storms is difficult and, often the outcome can become unaffordable. For example, Omaha Beach seawall in New Zealand was designed to prevent erosion from everyday waves only, and when a storm in 1976 carved out 10m behind the existing seawall the whole structure was destroyed (GeoResources, 2001).

Other Issues

Some further issues include: lack of long term trend data of seawall effects due to a relatively short duration of data records; modelling limitations and comparisons of different projects and their effects being invalid or unequal due to different beach types; materials; currents; and environments (Christchurch City Council, 2009).

Historical Examples

Seawall construction has existed since ancient times. In the 1st century BCE, Romans built a seawall / breakwater at Caesarea Maritima creating an artificial harbor (Sebastos Harbor). The construction used Pozzolana concrete which hardens in contact with sea water. Barges were constructed and filled with the concrete. They were floated into position and sunk. The resulting harbor / breakwater / sea wall is still in existence today - more than 2000 years later.

More recently, sea walls were constructed in 1623 in Canvey Island, UK, when great floods of the Thames estuary occurred, prompting the construction of protection for further events in this flood prone area (Council of Europe, 1999). Since then, seawall design has become more complex and intricate in response to an improvement in materials, technology and an understanding of how coastal processes operate. This section will outline some key case studies of seawalls in chronological order and describe how they have performed in response to tsunami or ongoing natural processes and how effective they were in these situations. Analysing the successes and shortcomings of seawalls during severe natural events allows their weaknesses to be exposed, and areas become visible for future improvement.

Pondicherry

On December 26, 2004, towering waves of the 2004 Indian Ocean earthquake tsunami crashed against India's south-eastern coastline killing thousands. However, the former French colonial enclave of Pondicherry (now Pondicherry) escaped unscathed. This was primarily due to French engineers who had constructed (and maintained) a massive stone seawall during the time which the city was a French colony. This 300 year old seawall effectively kept Pondicherry's historic centre dry even though tsunami waves drove water 24 feet above the normal high-tide mark.

The barrier was initially completed in 1735 and over the years, the French continued to fortify the wall, piling huge boulders along its 1.25 mile (2 km) coastline to stop erosion from the waves pounding the harbour. At its highest, the barrier running along the water's edge reaches about 27 feet above sea level. The boulders, some weighing up to a ton, are weathered black and brown. The sea wall is inspected every year and whenever gaps appear or the stones sink into the sand, the government adds more boulders to keep it strong (Allsop, 2002).

The Union Territory of Pondicherry recorded some 600 deaths from the huge tsunami waves that struck India's coast after the mammoth underwater earthquake (which measured 9.0 on the moment magnitude scale) off Indonesia, but most of those killed were fishermen who lived in villages beyond the artificial barrier which reinforces the effectiveness of seawalls.

Vancouver

The Vancouver Seawall is a stone seawall constructed around the perimeter of Stanley Park in Vancouver. The seawall was constructed initially as waves created by ships passing through the First Narrows were eroding the area between Prospect Point and Brockton Point. The Vancouver Seawall also exemplifies how seawalls can be utilised and valued for recreational activities and coastal sightseeing. A pedestrian, cycling and roller blading pathway exists on the seawall and has been extended far outside the parameters of Stanley Park. Construction of the seawall began in 1917, and since then this pathway has become one of the most used features of the park by both locals and tourists and now extends 22 km in total (Belyea & Ross, 1992). The construction of the seawall also provided employment for relief workers during the Great Depression and seamen from HMCS *Discovery* on Deadman's Island who were facing punishment detail in the 1950s (Steele, 1985).

Overall, the Vancouver Seawall is a prime example of how seawalls can simultaneously provide shoreline protection and a source of recreation which enhances human enjoyment of the coastal environment. It also illustrates that although shoreline erosion is a natural process, human activities, interactions with the coast and poorly planned shoreline development projects can accelerate natural erosion rates.

Japan

At least 43 percent of Japan's 29,751 km kilometre coastline is lined with concrete seawalls or other structures designed to protect the country against high waves, typhoons or even tsunamis (New York Times, 2011). When a Tsunami struck in 2011 following a magnitude 9 offshore earthquake, the seawalls in most areas were overwhelmed. In Kamaishi, 4-metre waves surmounted the seawall —the world's largest, erected a few years ago in the city's harbour at a depth of 63 metres, a length of 2 kilometres and a cost of $1.5 billion — and eventually submerged the city centre (Musubi, 2011).

The risks of dependence on seawalls was most evident in the crisis at the Dai-ichi and Dai-ni nuclear power plants, both located along the coast close to the earthquake zone, as the tsunami washed over walls that were supposed to protect the plants. Arguably, the additional defence provided by the seawalls presented an extra margin of time for citizens to evacuate and also stopped some of the full force of energy which would have caused the wave to climb higher in the backs of coastal valleys. In contrast, the seawalls also acted in a negative way to trap water and delay its retreat.

The failure of the world's largest seawall, which cost $1.5 billion to construct, shows that building stronger sea walls to protect larger areas would have been too costly to be effective. In the case of the ongoing crisis at the nuclear power plants, higher and stronger sea walls should have been built if power plants were to be built at that site. Fundamentally, the devastation in coastal areas and a final death toll predicted to exceed 10,000 could push Japan to redesign its seawalls or consider more effective alternative methods of coastal protection for extreme events. Such hardened coastlines can also provide a false sense of security to property owners and local residents as evident in this situation (Msubi, 2011).

Revetment

In stream restoration, river engineering or coastal management, revetments are sloping structures placed on banks or cliffs in such a way as to absorb the energy of incoming water. In military engineering they are structures, again sloped, formed to secure an area from artillery, bombing, or stored explosives. River or coastal revetments are usually built to preserve the existing uses of the shoreline and to protect the slope, as defence against erosion.

Freshwater Revetments

Many revetments are used to line the banks of freshwater rivers, lakes, and man-made reservoirs, especially to prevent damage during periods of floods or heavy seasonal rains. Many materials may be used: wooden piles, loose-piled boulders or concrete shapes, or more solid banks.

Revetments as Coastal Defence

Revetments are used as a low-cost solution for coastal erosion defence in areas where crashing waves may otherwise deplete the coastline. Wooden revetments are made of planks laid against wooden frames so that they disrupt the force of the water. Although once popular, the use of wooden revetments has largely been replaced by modern concrete-based defence structures such as tetrapods.

Tetrapods

In coastal engineering, a tetrapod is a four-legged concrete structure used as armour unit on breakwaters. The tetrapod's shape is designed to dissipate the force of incoming waves by allowing water to flow around rather than against it, and to reduce displacement by allowing a random distribution of tetrapods to mutually interlock.

Fortifications

According to the U.S. National Park Service, and referring mostly to their employment in the American Civil War, a revetment is defined as a "retaining wall constructed to support the interior slope of a parapet. Made of logs, wood planks, fence rails, fascines, gabions, hurdles, sods, or stones, the revetment provided additional protection from enemy fire, and, most importantly, kept the interior slope nearly vertical. Stone revetments commonly survive. A few log revetments have been preserved due to high resin pine or cypress and porous sandy soils. After an entrenchment was abandoned, many log or rail revetments were scavenged for other uses, causing the interior slope to slump more quickly. An interior slope will appear more vertical if the parapet eroded with the revetment still in place."

Riprap

Riprap, also known as rip rap, rip-rap, shot rock, rock armour or rubble, is rock or other material used to armor shorelines, streambeds, bridge abutments, pilings and other shoreline structures against scour, water or ice erosion. It is made from a variety of rock types, commonly granite or limestone, and occasionally concrete rubble from building and paving demolition. It can be used on any waterway or water containment where there is potential for water erosion.

Protection Mechanism

Riprap works by absorbing and deflecting the energy of waves before they reach the defended structure. The size and mass of the riprap material absorbs the impact energy of waves, while the gaps between the rocks trap and slow the flow of water, lessening its ability to erode soil or structures on the coast. The mass of riprap also provides protection against impact damage by ice or debris, which is particularly desirable for bridge supports and pilings. It is frequently used to protect the base of old Edwardian and Victorian sea walls, which being vertical are often undermined. The riprap absorbs the impact of the waves as they shoot up the wall and then fall back down.

Gabion

A gabion (from Italian *gabbione* meaning "big cage"; from Italian *gabbia* and Latin *cavea* meaning "cage") is a cage, cylinder, or box filled with rocks, concrete, or sometimes sand and soil for use in civil engineering, road building, and military applications. For erosion control, caged riprap is used. For dams or in foundation construction, cylindrical metal structures are used. In a military context, earth- or sand-filled gabions are used to protect artillery crews from enemy fire.

Leonardo da Vinci designed a type of gabion called a *Corbeille Leonard* ("Leonard[o] basket") for the foundations of the San Marco Castle in Milan.

Civil Engineering

The most common civil engineering use of gabions is to stabilize shorelines, streambanks or slopes against erosion. Other uses include retaining walls, temporary floodwalls, silt filtration from runoff, for small or temporary/permanent dams, river training, or channel lining. They may be used to direct the force of a flow of flood water around a vulnerable structure. Gabions are also used as fish barriers on small streams. A gabion wall is a retaining wall made of stacked stone-filled gabions tied together with wire. Gabion walls are usually *battered* (angled back towards the slope), or stepped back with the slope rather than stacked vertically.

Gabion baskets have some advantages over loose riprap because of their modularity and ability to be stacked in various shapes; they are also resistant to being washed away by moving water. Gabions also have advantages over more rigid structures because they can conform to ground movement, dissipate energy from flowing water, and drain freely. Their strength and effectiveness may increase with time in some cases, as silt and vegetation fill the interstitial voids and reinforce the structure. They are sometimes used to keep stones which may fall from a cutting or cliff from endangering traffic on a thoroughfare.

The life expectancy of gabions depends on the lifespan of the wire, not on the contents of the basket. The structure will fail when the wire fails.

Galvanized steel wire is most common, but PVC-coated and stainless steel wire are also used. PVC-coated galvanized gabions have been estimated to survive for 60 years. Some gabion manufacturers guarantee a structural consistency of 50 years.

In the United States, gabion use within streams first began with projects completed from 1957-1965 on North River, Virginia and Zealand River, NH. More than 150 grade-control structures, bank revetments and channel deflectors were constructed on the two U.S. Forest Service sites.

Eventually, a large portion of the instream structures failed due to undermining and lack of structural integrity of the baskets. In particular, corrosion and abrasion of wires by bedload movement compromised the structures, which then sagged and collapsed into the channels. Other gabions were toppled into channels as trees grew and enlarged on top of gabion revetments leveraging them towards the river channels.

Gabions have also been used in building, as in the Dominus Winery in Napa Valley, California. The exterior is formed by modular wire mesh gabions containing locally quarried stone; this construction creates an environment of moderate temperatures within the building.

Variations in Design

There are various special designs of gabions to meet particular functional requirements and some special terms for particular forms have come into use. For example:

- Bastion: a gabion lined internally with a membrane, typically of nonwoven geotextile to permit use of a granular soil fill instead of rock.
- Mattress: a form of gabion with relatively small height relative to the lateral dimensions; commonly very wide. For protecting surfaces from wave erosion and similar attack, rather than building or supporting high structures.
- Trapion: a form of gabion with a trapezoidal cross section, designed for stacking to give a face that is sloping rather than stepped. The term is in wide usage, but in contexts related to gabions at least, appears to be a trademark registered by Betafence Limited.

Breakwater (Structure)

Breakwaters are structures constructed on coasts as part of coastal defence or to protect an anchorage from the effects of both weather and longshore drift.

Purposes of Breakwaters

Offshore breakwaters, also called bulkhead, reduce the intensity of wave action in inshore waters and thereby reduce coastal erosion or provide safe harborage. Breakwaters may also be small structures designed to protect a gently sloping beach and placed one to three hundred feet offshore in relatively shallow water.

An anchorage is only safe if ships anchored there are protected from the force of high winds and powerful waves by some large underwater barrier which they can shelter behind. Natural harbours are formed by such barriers as headlands or reefs. Artificial harbours can be created with the help of breakwaters. Mobile harbours, such as the D-Day Mulberry harbours, were floated into position and acted as breakwaters. Some natural harbours, such as those in Plymouth Sound, Portland Harbour and Cherbourg, have been enhanced or extended by breakwaters made of rock.

Unintended Consequences

The dissipation of energy and relative calm water created in the lee of the breakwaters often encourage accretion of sediment (as per the design of the breakwater scheme). However this can lead to excessive salient build up, leading to tombolo formation reducing longshore drift shoreward of the breakwaters (Sea Palling, UK). This trapping of sediment can cause adverse effects down drift of the breakwaters leading to beach sediment starvation and increased erosion. This may then lead to further engineering protection being needed down drift of the breakwater development.

Breakwaters are subject to damage, and overtopping in severe storms events.

Construction

Breakwaters can be constructed with one end linked to the shore, in which case they are usually classified as sea walls; otherwise they are positioned offshore from as little as 100m up to 300-600m from the original shoreline. There are two main types of offshore breakwater, single and multiple; single as the name suggests means the breakwater consists of one unbroken barrier, which multiple breakwaters (in numbers anywhere from 2-20) are positioned with gaps in between (50-300m). Length of gap is largely governed by the interacting wavelengths.

Breakwaters may be either fixed or floating, and impermeable or permeable to allow sediment transfer shoreward of the structures, the choice depending tidal range and water depth. They usually consist of large pieces of rock (granite) weighting up to 16 tonnes each or rubble-mound. Their design is influenced by the angle of wave approach and other environmental parameters. Breakwater construction can be either parallel or perpendicular to the coast, depending on the shoreline requirements.

Types of Breakwater Structures

A breakwater structure is designed to absorb the energy of the waves that hit it, either by using mass (e.g., with caissons), or by using a revetment slope (e.g., with rock or concrete armour units).

In Coastal Engineering, a revetment is a land backed structure whilst a breakwater is a sea backed structure (i.e., water on both sides).

Caisson breakwaters typically have vertical sides and are usually erected where it is desirable to berth one or more vessels on the inner face of the breakwater. They use the mass of the caisson and the fill within it to resist the overturning forces applied by waves hitting them. They are relatively expensive to construct in shallow water, but in deeper sites they can offer a significant saving over revetment breakwaters.

Rubble mound breakwaters use structural voids to dissipate the wave energy. Rock or concrete armour units on the outside of the structure absorb most of the energy, while gravels or sands prevent the wave energy's continuing through the breakwater core. The slopes of the revetment are typically between 1:1 and 1:2, depending upon the materials used. In shallow water, revetment breakwaters are usually relatively inexpensive. As water depth increases, the material requirements, and hence costs, increase significantly.

Advanced Numerical Study

The Maritime Engineering Division of the University of Salerno (MEDUS) developed a new procedure for studying in greater detail the interactions between maritime breakwaters (submerged or emerged) and the waves that hit them by making integrated use of CAD and CFD software.

In the numerical simulations, the filtration motion of the fluid within the interstices, which normally exist in a breakwater, is estimated by integrating the RANS equations, coupled with a RNG turbulence model inside the voids, instead of using classical equations for porous media. The breakwaters were modelled, in analogy to full size construction or physical laboratory tests, by overlapping three-dimensional elements and having the numerical grid thickened in order to have some computational nodes along the flow paths among the breakwater's blocks.

Cliff Stabilization

Cliff stabilization is a coastal management erosion control technique. This is most suitable for softer or less stable cliffs. Generally speaking, the cliffs are stabilised through dewatering (drainage of excess rainwater to reduce water-logging) or anchoring (the use of terracing, planting, or wiring to hold cliffs in place).

Entrance Training Walls

Rock or concrete walls built to constrain a river or creek discharging across a sandy coastline. The walls help to stabilise and deepen the channel which benefits navigation, flood management, river erosion and water quality but can cause coastal erosion due to the interruption of longshore drift. One solution is the installation of a sand bypassing system to pump sand under and around the entrance training walls.

Cost – Expensive - Gold Coast Seaway was a A$50M project in the 1980s and the adjacent sand bypassing project costs A$3M per year to pump 500,000 cubic metres of sand across the trained entrance.

Floodgates

Storm surge barriers, or floodgates, were introduced after the North Sea Flood of 1953 and are a prophylactic method to prevent damage from storm surges or any other type of natural disaster that could harm the area they "protect". They are habitually open and allow free passage, but close when the land is under threat of a storm surge. The Thames Barrier is an example of such a structure.

4

Soft Engineering Methods

Beach Nourishment

Beach nourishment— also referred to as beach replenishment or sand replenishment —describes a process by which sediment (usually sand) lost through longshore drift or erosion is replaced from sources outside of the eroding beach. A wider beach can reduce storm damage to coastal structures by dissipating energy across the surf zone, protecting upland structures and infrastructure from storm surges, tsunamis and unusually high tides. Beach nourishment is typically part of a larger coastal defence scheme. Nourishment is typically a repetitive process, since it does not remove the physical forces that cause erosion, but simply mitigates their effects.

The first nourishment project in the U.S. was at Coney Island, New York in 1922-23 and is now a common shore protection measure utilized by public and private entities. Nourishment is one of three commonly accepted methods for protecting shorelines. The structural alternative involves constructing a seawall, revetment, groin or breakwater. Alternatively, with "managed retreat" the shoreline is left to erode, while relocating buildings and infrastructure further inland. Nourishment gained popularity because it preserved beach resources and avoided the negative effects of hard structures. Instead, nourishment creates a "soft" (i.e., non-permanent) structure by creating a larger sand reservoir, pushing the shoreline seaward.

Causes of Erosion

Beaches can erode both naturally and due to impact of humans. Erosion is a natural response to storm activity. During storms, sand from the visible beach submerges to form storm bars that protect the beach. Submersion is only part of the cycle. During calm weather smaller waves return sand from the storm bar to the visible beach surface in a process called accretion. The term *erosion* conjures visions of environmental damage so the term submersion often replaces it in describing a healthy sandy beach. Continental drift erodes coastlines naturally. Ocean currents can change, disrupting the submersion/ accretion cycle.

Some beaches do not have enough sand available to coastal processes to respond naturally to storms. When there is not enough sand left available on a beach, then there is no recovery of the beach following storms. Many areas of high erosion are due to human activities. Reasons can include seawalls locking up sand dunes, coastal structures like ports and harbours that prevent longshore drift, dams and other river management structures. These activities interfere with the natural sediment flows either through dam construction (thereby reducing riverine sediment sources) or construction of littoral barriers such as jetties, or by deepening of inlets; thus preventing longshore transport of sediment across these channels.

Visible and Submerged Sand

The distinction between total sand in a beach and the proportion of the sand above the waterline (submersion fraction) critically impacts beach nourishment. Two beaches with the same amount of visible sand may look much different under water. An eroded beach with substantial submerged sand surrounding it may recover without nourishment. Nourishing a beach that has little submerged sand requires addressing the reason that the submerged sand is missing. Otherwise, the same forces that stripped the submerged sand once are likely to do so again. The amount of submerged sand eroded is typically much greater than the amount of missing sand on shore. Replacing only the visible sand is insufficient without replacing the sand that once supported the accretion part of the process. In that circumstance, the beach is unstable and the visible sand quickly erodes. If human activity is a major cause of the erosion, mitigating that activity may be more cost effective over both short and long term periods than nourishment.

Requirements for Effective Nourishment

Sediment texture (grain size and sorting) is critical for success. Sand fill must be compatible with native beach sand. In some cases, beaches have been nourished using a finer sand than the original.

Thermoluminescence monitoring reveals that storms can erode such beaches far more quickly than the natural beach. This was observed at the Waikiki nourishment project in Hawaii.

Profile Nourishment

Beach Profile Nourishment describes programmes that nourish the full beach profile. In this instance, "profile" means the slope of the uneroded beach from above the water to well out to sea, not just the visible portion. The Gold Coast profile nourishment programme placed 75% of its total sand volume below low water level. Some coastal authorities *overnourish* the below water beach (aka "nearshore nourishment") so that over time the natural beach increases in size. These approaches do not permanently protect beaches eroded by human activity. Doing so still requires mitigating that activity.

Replenishment Material

The selection of suitable material for a particular project depends upon the design needs, environmental factors, transport costs considering both short and long-term implications.

Type	*Description*	*Environmental issues*
Offshore	Exposure to open sea makes this the most difficult operational environment. Must consider the effects of altering depth on wave energy at the shoreline. May be combined with a navigation project.	Impacts on hard bottom and migratory species.
Inlet	Sand between jetties in a stabilized inlet. Often associated with dredging of navigational channels and the ebb- or flood-tide deltas of both natural and jettied inlets.	
Accretional Beach	Generally not suitable because of damage to source beach.	
Upland	Generally the easiest to obtain permits and assess impacts from a land source. Offers opportunities for mitigation. Limited quantity and quality of economical deposits.	Potential secondary impacts from mining and overland transport.
Riverine	Potentially high quality and sizeable quantity. Transport distance a possible cost factor.	May interrupt natural coastal sand supply.
Lagoon	Often excessively fine grained. Often close to barrier beaches and in sheltered waters, easing construction. Principal sources are flood-tide deltas.	Can compromise wetlands.
Artificial or non-indigenous	Typically, high transport and redistribution costs. Some laboratory experiments done on recycling broken glass. Aragonite from Bahamas a possible source.	
Emergency	Deposits near inlets and local sinks and sand from stable beaches with adequate supply. Generally used only following a storm or given no other affordable option. May be combined with a navigation project.	Harm to source site. Poor match to target requirements.

The most important material characteristic is the sediment's grain size, which must closely match the native material. Excess silt and clay fraction (mud) versus the natural turbidity in the nourishment area disqualifies some materials. Projects that did not match grain sizes performed relatively poorly. Nourishment sand that is only slightly smaller than native sand can result in significantly narrower equilibrated dry beach widths compared to sand the same size as (or larger than) native sand. Evaluating material fit requires a sand survey that usually includes geophysical profiles and surface and core samples.

'*Advantages:*

1. widens the recreational beach.
2. Structures behind beach are protected as long as the added sand remains.

Disadvantages:

1. Beach nourishment sand often (in fact, usually) erodes faster than the natural sand on the beach.
2. Beach nourishment is expensive, and must be repeated periodically.
3. The beach turns into a construction zone during nourishment.
4. The process of nourishment may damage, destroy or otherwise hurt marine and beach life by burying it.
5. The sand added to the beach is often different from the natural beach sand. It can be hard to find a perfect match.

Environmental Issues

Beach nourishment has significant impacts on local ecosystems. Nourishment may cause direct mortality to sessile organisms in the target area by burying them in the new sand. Seafloor habitat in both source and target areas are disrupted, e.g., when sand is deposited on coral reefs or when deposited sand hardens. Imported sand may differ in character (chemical makeup, grain size, non-native species) from that of the target environment. Reduced light availability, affecting nearby reefs and submerged aquatic vegetation. Imported sand may contain material toxic to local species. Removing material from near-shore environments may destabilize the shoreline, in part by steepening its submerged slope. Related attempts to reduce future erosion may provide a false sense of security that increases development pressure.

Sea Turtles

Newly-deposited sand can harden and complicate nest-digging for turtles. However, nourishment can provide more/better habitat for them, as well as for sea birds and beach flora. Florida addressed the concern that dredge pipes would suck turtles into the pumps by adding a special grill to the dredge pipes that ensured that turtles were not sucked in with the sand.

Alternatives/Complements to Nourishment

Nourishment is not the only technique used to address eroding beaches. Others can be used singly or in combination with nourishment, driven by economic, environmental, and political considerations.

Human activities such as dam construction can interfere with natural sediment flows (thereby reducing riverine sediment sources.) Construction of littoral barriers such as jetties and deepening of inlets can prevent longshore sediment transport.

Structural

The structural approach attempts to prevent erosion. Armoring involves building revetments, seawalls, detached breakwaters, groins, etc. Structures that run parallel to the shore (seawalls or revetments) prevent erosion. While this protects structures, it doesn't not protect the beach which is outside the wall. The beach disappears over a period that ranges from months to decades.

Groynes and breakwaters that run perpendicular to the shore protect the shore from erosion. Filling a breakwater with imported sand can stop the breakwater from trapping sand from the littoral stream (the ocean running along the shore.) Otherwise the breakwater may deprive downstream beaches of sand and accelerate erosion there.

Armoring may restrict beach/ocean access, enhance erosion of adjacent shorelines, and requires long-term maintenance.

Managed Retreat

A second option is to move structures and infrastructure inland as the shoreline erodes. Retreat is more often chosen in areas of rapid erosion and in the presence of little or obsolete development.

Recruitment

Appropriately constructed and sited fences can capture blowing sand, building/restoring sand dunes, and progressively protecting the beach from the wind, and the shore from blowing sand.

Beach Drainage

All beaches lose and gain sand depending on the tides, precipitation, wind, waves etc. Wet beaches tend to lose sand. Waves infiltrate dry beaches easily and deposit sandy sediment. Generally a beach is wet during falling tide, because the sea sinks faster than the beach drains. As a result most erosion happens during falling tide. Beach drainage (beach dewatering) using Pressure Equalizing Modules (PEMs) allow the beach to drain more effectively during falling tide. Fewer hours of wet beach translate to less erosion.

Permeable PEM tubes inserted vertically into the foreshore connect the different layers of groundwater. The groundwater enters the PEM tube allowing gravity to conduct it to a coarser sand layer, where it can drain more quickly. The PEM modules are placed in a row from the dune to the mean low waterline.

Distance between rows is typically 300 feet (91 m) but this is project-specific. PEM systems come in different sizes. Modules connect layers with varying hydraulic conductivity. Air/water can enter and equalize pressure.

PEMs are minimially invasive, typically covering approximately 0.00005% of the beach. The effects are local, and bring no known harm to flora or fauna, including nesting sea turtles. The tubes are below the beach surface, with no visible impact. Installation takes little time. Costs are low, because PEMs require few materials, and no energy-intensive dredging. PEM installations have been installed on beaches in Denmark, Sweden, Malaysia, and Florida, USA. The effectiveness of beach dewatering, however, is debatable and has not been proven convincingly on life-size beaches.

Costs

Nourishment is typically a repetitive process, since nourishment does not remove the physical forces that cause erosion; it simply mitigates their effects. A benign environment increases the interval between nourishment projects, reducing costs. Conversely, high erosion rates may render nourishment financially impractical.

In many coastal areas, the economic impacts of a wide beach can be substantial. The 10 miles (16 km)–long shoreline fronting Miami Beach, Florida was replenished over the period 1976–1981. The project cost approximately $64,000,000 and revitalized the area's economy. Prior to nourishment, in many places the beach was too narrow to walk along, especially during high tide.

History

The first nourishment project in the U.S. was constructed at Coney Island, New York in 1922-1923.

Nourishment Projects

The setting of a beach nourishment project is key to design and potential performance. Possible settings include a long straight beach, an inlet that may be either natural or modified, and a pocket beach. Rocky or seawalled shorelines, that otherwise have no sediment, present unique problems.

Cancun, Mexico

Federal and state governments in Mexico have invested about $71 million ($957 million pesos) throughout the state of Quintana Roo in restoring the beaches along Cancun, Playa del Carmen, and Cozumel.

Hurricane Wilma hit the beaches of Cancun and the Riviera Maya in 2005. The initial nourishment project was unsuccessful, leading to a second round that began in September 2009, and was scheduled to complete in early 2010. The project designers and the government committed to invest in beach maintenance to address future erosion. Project designers considered factors such as the time of year and sand characteristics such as density. Restoration in Cancun was expected to deliver 1.3 billion US gallons (4,900,000 m^3) of sand to replenish 450 metres (1,480 ft) of coastline. This time, the beach is promised to last at least 10 years.

Northern Gold Coast, Queensland, Australia

Gold Coast beaches in Queensland, Australia have experienced periods of severe erosion. In 1967 a series of 11 cyclones removed most of the sand from Gold Coast beaches. The Government of Queensland engaged engineers from Delft University in the Netherlands to advise them. The 1971 Delft Report outlined a series of works for Gold Coast Beaches, including beach nourishment and an artificial reef. By 2005 most of the recommendations had been implemented.

The Northern Gold Coast Beach Protection Strategy (NGCBPS) was a A$10 million investment. NGCBPS was developed between 1992 and 1999 and the works were completed between 1999 and 2003. The project included dredging 3,500,000 cubic metres (4,600,000 cu yd) of compatible sand from the Gold Coast Broadwater and delivering it through a pipeline to nourish 5 kilometres (3.1 mi) between of beach between Surfers Paradise and Main Beach. The new sand was stabilized by an artificial reef constructed at Narrowneck out of huge geotextile sand bags. The new reef was designed to improve wave conditions for surfing. A key monitoring programme for the NGCBPS is the ARGUS coastal camera system operated by the University of New South Wales.

The cost/benefit ratio for NGCBPS was conservatively estimated at 75:1 for a A$10 million investment into beach replenishment. The benefits were estimated from a model of lost visitor nights in hotels following previous erosion events. NGCBPS so improved beach health that recovery following minor and moderate storms occurred within weeks. Additional unquantified benefits included lifestyle benefits for residents, additional public open space and improved fishing, diving and surfing conditions.

Netherlands

More than one-quarter of the Netherlands is below sea level and about 81% of the coast consists of sand dune or beach. The shoreline is closely monitored by yearly recording of the cross section at points 250 metres (820 ft) apart, to ensure adequate protection. Where long-term erosion is identified, beach nourishment using high-capacity suction dredgers is deployed.

Hawaii

Waikiki: Hawaii planned to replenish Waikiki beach in 2010. Budgeted at $2.5 million, the project covered 1,700 feet (520 m) in an attempt to return the beach to its 1985 width. Prior opponents supported this project, because the sand was to come from nearby shoals, reopening a blocked channel and leaving the overall local sand volume unchanged, while closely matching the "new" sand to existing materials. The project planned to apply up to 24,000 cubic yards (18,000 m^3) of sand from deposits located 1,500 to 3,000 feet (460 to 910 m) offshore at a depth of 10 to 20 feet (3.0 to 6.1 m). The project was larger than the prior recycling effort in 2006-07, which moved 10,000 cubic yards (7,600 m^3).

Maui

Maui, Hawaii illustrated the complexities of even small-scale nourishment projects. A project at Sugar Cove transported upland sand to the beach. The sand allegedly was finer than the original sand and contained excess silt that enveloped coral heads, smothering the coral and killing small animals that lived in and around it. As in other projects, on-shore sand availability was limited, forcing consideration of more expensive offshore sources.

A second project, along Stable Road, that attempted to slow rather than halt erosion, was stopped halfway towards its goal of adding 10,000 cubic yards (7,600 m^3) of sand. The beaches had been retreating at a "comparatively fast rate" for half a century. The restoration was complicated by the presence of old seawalls, groins, piles of rocks and other structures.

This project used sand-filled Geotube groins placed that were originally to remain in place for up to 3 years. The pipeline was anchored by concrete blocks attahced by fibre straps. A video showed the blocks bouncing off the coral in the current, killing whatever they touched. In places the straps broke, allowing the pipe to move across the reef, "planing it down". Bad weather exacerbated the damaging movement and led to project shutdown. The smooth, cylindrical Geotubes could be difficult to climb over before they were covered by sand.

Supporters claimed that 2010's seasonal summer erosion was less than in prior years, although the beach was narrower after the restoration ended than in 2008. Authorities were studying whether to require the project to remove the groins immediately. Potential alternatives to Geotubes for moving sand included floating dredges and/or trucking in sand dredged offshore.

A final consideration was sea level rise and that Maui was sinking under its own weight. Both islands surround massive mountains (Haleakala, Mauna Loa, and Mauna Kea) and were expanding a giant dimple in the ocean floor, some 30,000 feet (9,100 m) below the mountain summits.

Hong Kong

The beach in Gold Coast was built as an artificial beach in the 1990s with HK$60m. Sands have to be supplied periodically, especially after typhoons, to keep the beach viable.

Measuring Project Impact

Nourishment projects usually involve physical, environmental and economic objectives. Typical physical measures include dry beach width, remaining post-storm sand volume, post-storm damage avoidance assessments and aqueous sand volume. Environmental measures include marine life distribution, habitat and population counts. Economic impacts include recreation, tourism, flood and "disaster" prevention. Techniques for incorporating nourishment projects into flood insurance costs and disaster assistance remain controversial.

The ability to predict the performance of a beach nourishment project is best for a project constructed on a long, straight shoreline without the complications of inlets or engineered structures. In addition, predictability is better for overall performance, e.g., average shoreline change, rather than shoreline change at a specific location.

Nourishment can affect eligibility in the National Flood Insurance Programme and federal disaster assistance.

Sand Dune Stabilization

Vegetation can be used to encourage dune growth by trapping and stabilising blown sand.

Cost – est. of £1.1 million per annum

Beach Drainage

Beach drainage or beach face dewatering lowers the water table locally beneath the beach face. This causes accretion of sand above the drainage system.

Grant (1946) – the elevation of the beach watertable had an important bearing on deposition and erosion across the foreshore. A high watertable coincided with periods of accelerated beach erosion, and conversely, a low watertable coincided with pronounced aggradation of the foreshore A lower watertable (unsaturated beach face) facilitates deposition by reducing flow velocities during backwash and prolonging laminar flow. In contrast, a high watertable results in condition favouring beach erosion. With the beach in a saturated state, Grant proposed that backwash velocity is accelerated by the addition of groundwater seepage out of the beach within the effluent zone.

Turner and Leatherman (1997) moving from the origins and development of the dewatering concept to field and laboratory studies available at the time of writing concluded that there was too little evidence for being convinced that the systems had a positive effect. None of the case studies provide full scientific evidence of indisputable positive results regarding beach stabilisation although in some cases an overall positive performance was reported. In many cases no adequate long-term monitoring was undertaken at a frequency high enough to discriminate the response to high energy erosive events.

A useful side effect of the system is that the collected seawater is very pure because of the sand filtration effect. It may be discharged back to sea but can also be used to oxygenate stagnant inland lagoons /marinas or used as feed for heat pumps, desalination plants, land-based aquaculture, aquariums or seawater swimming pools.

Beach drainage systems have been installed in many locations around the world to halt and reverse erosion trends in sand beaches. Twenty four beach drainage systems have been installed since 1981 in Denmark, USA, UK, Japan, Spain, Sweden, France, Italy and Malaysia.

Costs

The costs of installation and operation per metre of shoreline protection will vary due to

- system length (non-linear cost elements)
- pump flow rates (sand permeability, power costs)
- soil conditions (presence of rock or impermeable strata)
- discharge arrangement /filtered seawater utilization
- drainage design, materials selection & installation methods
- geographical considerations (location logistics)
- regional economic considerations (local capabilities /costs)
- study requirements /consent process.

The costs associated with a beach drainage system are generally considerably lower than hard engineered structures. They also compare very favourably with beach nourishment projects, particularly when long-term project economics are considered (nourishment projects often have a limited life or a programme of re-nourishment).

Monitoring Coastal Zones

Coastal zone managers are faced with difficult and complex choices about how best to reduce property damage in the shorelines. One of the problems they face is error and uncertainty in the information available to them on the processes that cause erosion of beaches. Video-based monitoring lets collect data continuously at low cost and produce analyses of shoreline processes over a wide range of averaging intervals.

Event Warning Systems

Event warning systems, such as tsunami warnings and storm surge warnings, can be used to minimize the human impact of catastrophic events that cause coastal erosion. Storm surge warnings can also be used to determine when to close floodgates to reduce the physical impact of such events.

Wireless sensor networks can be deployed quickly to set up a coastal erosion monitoring system, and scaled accordingly.

Shoreline Mapping

Defining the shoreline is a difficult task due to the dynamic nature of the coast and the intended application of the shoreline (Graham et al. 2003; Boak & Turner 2005). Given this idea the shoreline must therefore be considered in a temporal sense whereby the scale is dependent on the context of the investigation (Boak & Turner 2005). The following definition of the coast and shoreline is most commonly employed for the purposes of shoreline mapping. The coast comprises the interface between land and sea, and the shoreline is represented by the margin between the two (Woodroffe, 2002). Due to the dynamic nature of the shoreline coastal investigators adopt the use of shoreline indicators to represent the true shoreline position (Boak & Turner 2005).

Shoreline Indicator

The choice of shoreline indicator is a primary consideration in shoreline mapping. According to Leatherman (2003) it is important that indicators are easily identified in the field and on aerial photography. Shoreline indicators may be physical beach morphological features such as the berm crest, scarp edge, vegetation line, dune toe, dune crest and cliff or the bluff crest and toe. Alternatively, non-morphological features may also be used. These indicators are based on water level including the high water line, mean high water line, wet/dry boundary, and the physical water line (Pajak & Leatherman 2000).

The high water line (HWL), defined as the wet/dry line is the most commonly used shoreline indicator because it is visible in the field, and can be interpreted on both colour and grey scale aerial photographs (Leatherman, 2003; Crowell et al. 1991). The HWL represents the landward extent of the most recent high tide and is characterised by a change in sand colour due to repeated, periodic inundation by high tides. The HWL is portrayed on aerial photographs by the most landward change in colour or grey tone (Boak & Turner 2005).

Importance and Application

The location of the shoreline and its changing position over time is of fundamental importance to coastal scientists, engineers and managers (Boak & Turner 2005; Pajack & Leatherman 2002). Present day shoreline monitoring campaigns provide information about historic shoreline location and movement, and about predictions of future change (Appeaning Addo et al. 2008). More specifically the position of the shoreline in the past, at present and where it is predicted to be in the future is useful for in the design of coastal protection, to calibrate and verify numerical models to assess sea level rise, map hazard zones and formulate policies to regulate coastal development. Accurate and consistent delineation of the shoreline is integral to all of these tasks. The location of the shoreline also provides information regarding shoreline reorientation adjacent to structures, beach width, volume and rates of historical change (Boak & Turner 2005; Pajack & Leatherman 2002).

Data Sources

A variety of data sources are available for examining shoreline position however, the availability of historical data is limited at many coastal sites and so the choice of data source is largely limited to what is available for the site at a given time (Boak & Turner 2005). Shoreline mapping techniques applied to data sources have moved towards automation in association with technological advances and the need to reduce uncertainty. Although these changes have resulted in improvement in coastal data processing and storage capabilities, the frequent change in technology has prevented the emergence of one standard method of shoreline mapping. This has occurred because each data source and associated method have their own unique capabilities and shortcomings (Moore 2000). A number of the data sources used for shoreline mapping and their associated advantages and disadvantages are discussed below.

Historical Maps

In the event that a study requires the shoreline position to be mapped before the development of aerial photographs, or if the location has poor photograph coverage it is necessary to employ historical maps in order to detail shoreline position (Moore 2000). The main advantage and reason for using historical maps is that they are able to provide a historic record that is not available from other data sources. Many potential errors however are associated with historical coastal maps and charts. Such errors may be associated with scale, datum changes, distortions from uneven shrinkage, stretching, creases, tears and folds, different surveying standards, different publication standards, and projection errors (Boak & Turner 2005). The severity of these errors depends on the accuracy standards met by each map and the physical changes that have occurred since the publication of the map (Anders & Byrnes 1991). The oldest reliable source of shoreline data in the United States dates back to the early-to-mid-19th century and is the U.S Coast and Geodetic Survey/National Ocean Service T-sheets (Morton 1991). In the United Kingdom, many maps and charts were deemed to be inaccurate until around 1750. The founding of the Ordnance Survey in 1791 has since improved the accuracy of the mapping.

Aerial Photographs

Aerial photographs have been used since the 1920s to provide topographical information about an area. They are therefore a good database for compilation of historical shoreline change maps. Aerial photographs are the most commonly used data source in shoreline mapping because many coastal areas have extensive aerial photo coverage therefore providing a valuable record of shoreline position (Moore 2000). In general, aerial photographs provide good spatial coverage of the coast however temporal coverage is very much site specific depending on the flight path of the aeroplane. A second disadvantage associated with aerial photography is that the interpretation of the shoreline position is subjective given the dynamic nature of the coastal environment. This combined with various distortions inherent in aerial photographs can lead to significant error levels (Moore 2000). The minimisation of further errors is discussed below.

Object Space Displacements

Conditions outside of the camera can cause objects in an image to be displaced from their true ground position. Such conditions may include ground relief, camera tilt and atmospheric refraction.

Relief displacement is prominent when photographing a variety of elevations. This situation causes objects above ground level to be displaced outward from the centre of the photograph and objects below ground level to be displaced towards the centre of the image. The severity of the displacement is affected negatively with decreases in flight altitude and as radial distance from the centre of the photograph increases. This distortion can be minimised by photographing numerous swaths and creating a mosaic of the images. This technique will create a focus for the centre of each photograph where distortion is minimised. It is important to note that this error is not common in shoreline mapping is the relief is fairly constant. It is however important to consider when mapping cliffs (Moore 2000).

Ideally aerial photographs are taken so the optical axis of the camera is perfectly perpendicular to the ground surface thereby creating a vertical photograph. Unfortunately this is not often the case and virtually all aerial photographs experience tilt whereby up to 3° is not uncommon (Camfield et al. 1996). In this situation the scale of the image will be larger on the upward side of the tilt axis and smaller on the downward side. Moore, (2000) notes that many coastal researchers have not realised the severity of this error and therefore do not consider it in their methods.

Radial Lens Distortion

Lens distortion varies as a function of radial distance from the iso-centre of the photograph meaning that the centre of the image is relatively distortion free, but as the angle of view increases the distortion becomes more prominent. This is a significant source of error in earlier aerial photography but as technology has increased and camera lens have become more refined it has become less of an issue with later photographs. Such a distortion is impossible to correct for without knowing the make and model of the lens used to capture the image. However if overlapping images have been acquired one can digitize the centre portions of the aerial photographs (Crowell et al. 1991).

Delineation of the Shoreline

The dynamic nature of the coast has meant that accurate mapping of an instantaneous shoreline position has been associated with significant uncertainty. This uncertainty arises because at any given time the position of the shoreline is influenced by the short-term effect of the tide and a wide variety of long term effects such as relative sea-level rise and along shore littoral sediment movement. Not only does this affect the accuracy of computed historic shoreline position but also any predicted future positions (Appeaning Addo et al. 2008). As mentioned earlier the HWL is most commonly used as a shoreline indicator. This can usually be seen as a significant tonal change on aerial photographs. There are however many errors associated with using the wet/dry line as a proxy for the HWL and shoreline. The errors of largest concern are the short term migration of the wet/dry line, interpretation of the wet/dry line on a photograph and measurement of the interpreted line position (Leatherman 2003; Moore 2000). Systematic errors such as the migration of the wet/dry line may arise from tidal and seasonal changes. Storm-induced erosion is another factor which may cause the wet/dry line to migrate landward. Field investigations have shown that these changes can be minimised by using only summertime data (Moore 2000; Leatherman 2003). Furthermore, the error bar can be significantly reduced by using the longest record of reliable data to calculate erosion rates (Leatherman

2003). Finally it is important to note that errors may arise due to the difficulty of measuring a single line on a photograph. For example where the pen line is 0.13 mm thick this translates to an error of ±2.6 m on a 1:20000 scale photograph.

Beach Profiling Surveys

Beach profiling surveys are typically repeated at regular intervals along the coast in order to measure short-term (daily to annual) variations in shoreline position and beach volume. (Smith & Zarillo 1990). Beach profiling is a very accurate source of information however measurements are generally subject to the limitations of conventional surveying techniques. Shoreline data derived from beach profiling is often spatially and temporally limited due to the high cost associated with such a labour intensive activity.

Shorelines are generally derived by interpolating between a series of discrete beach profiles. It is important to note however that the distance between the profiles is usually quite large and so the accuracy of the interpolating becomes compromised. In contrast to aerial photographs, survey data is limited to smaller lengths of shoreline generally less than ten kilometres (Boak & Turner 2005). Beach profiling data is commonly available in from regional councils in New Zealand such as those compiled by the Hawkes Bay Regional Council.

Remote Sensing

Technological advancement over the last decade has led to the development of a range of airborne, satellite and land based remote sensing techniques (Smith & Zarillo 1990). Some of the remotely sensed data sources are listed below:

- Multispectral and hyperspectral imaging
- Microwave sensors
- Global positioning system (GPS)
- Airborne light detection and ranging technology (LIDAR)

Remote sensing techniques are attractive as they are cost effective, reduce manual error and remove the subjective approach of conventional field techniques (Maiti et al. 2009).Remote sensing is a relatively new concept and so extensive historical observations are unavailable. Given this idea, it is important that coastal morphology observations are quantified by coupling remotely sensed data with other sources of information detailing historic shoreline position from archived sources (Appeaning Addo et al. 2008).

Video Analysis

Video analysis provides quantitative, cost-effective, continuous and long-term monitoring beaches (Turner et al. 2004). The advancement of coastal video systems over the past 15 years has resulted in the extraction of large amounts of geophysical data from images. Such data includes that about coastal morphology, surface currents and wave parameters. The main advantage of video analysis lies in the ability to reliably quantify these parameters with high resolution and coverage in both space and time.

This in particular highlights their potential importance as an effective coastal monitoring system and an aid to coastal zone management (Van Koningsveld et al. 2007). Interesting case studies have been carried out using video analysis. Turner et al. (2004) used a video-based ARGUS coastal imaging system to monitor and quantify the regional-scale coastal response to sand nourishment and construction of the world-first Gold Coast artificial (surfing) reef in Australia. In addition, Smit et al. (2007) demonstrated the added value of high resolution video observations for making short-term predictions of near shore hydrodynamic and morphological processes, at temporal scales of metres to kilometres and days to seasons.

Integrated Coastal Zone Management

Integrated coastal zone management (ICZM) or Integrated coastal management (ICM) is a process for the management of the coast using an integrated approach, regarding all aspects of the coastal zone, including geographical and political boundaries, in an attempt to achieve sustainability.

The European Commission defines the ICZM as follows:-

ICZM is a dynamic, multidisciplinary and iterative process to promote sustainable management of coastal zones. It covers the full cycle of information collection, planning (in its broadest sense), decision making, management and monitoring of implementation. ICZM uses the informed participation and cooperation of all stakeholders to assess the societal goals in a given coastal area, and to take actions towards meeting these objectives. ICZM seeks, over the long-term, to balance environmental, economic, social, cultural and recreational objectives, all within the limits set by natural dynamics. 'Integrated' in ICZM refers to the integration of objectives and also to the integration of the many instruments needed to meet these objectives. It means integration of all relevant policy areas, sectors, and levels of administration. It means integration of the terrestrial and marine components of the target territory, in both time and space. To further understand the idea of ICZM several aspects can be defined and further explained. The coastal zone, the concept of sustainability and the term integration all within a coastal management context can be individually defined, while the expectations and framework of ICZM can be further explained. This entry uses the example of the New Zealand national framework to illustrate ICZM.

Defining the Coastal Zone

Defining the Coastal zone is of particular importance to the idea of ICZM. But the fuzziness of borders due to the dynamic nature of the coast makes it difficult to clearly define. Most simply the coast can be thought of as an area of interaction between the land and the ocean. Ketchum (1972) defined the area as: The band of dry land and adjacent ocean space (water and submerged land) in which terrestrial processes and land uses directly affect oceanic processes and uses, and vice versa. Issues arise with the diversity of features present on the coast and the spatial scales of the interacting systems. Coasts being dynamic in nature are influenced differently all around the world. Influences such as river systems, may reach far inland increasing the complexity and scale of the zone. These issues make it difficult to clearly identify hinterlands and subscribe any subsequent management.

Whilst acknowledging a physical coastal zone, the inclusion of ecosystems, resources and human activity within the zone is important. It is the human activities that warrant management. These activities are responsible for disrupting the natural coastal systems. To add to the complexity of this zone, administrative boundaries use arbitrary lines that dissect the zone, often leading to fragmented management. This sectored approach focuses on specific activities such as land use and fisheries, often leading to adverse effects in another sector.

The Importance of the Coastal Zone and the Need for Management

The dynamic processes that occur within the coastal zones produce diverse and productive ecosystems which have been of great importance historically for human populations. Coastal margins equate to only 8% of the worlds surface area but provide 25% of global productivity. Stress on this environment comes with approximately 70% of the world's population being within a day's walk of the coast. Two-thirds of the world's cities occur on the coast.

Valuable resources such as fish and minerals are considered to be common property and are in high demand for coastal dwellers for subsistence use, recreation and economic development. Through the perception of common property, these resources have been subjected to intensive and specific exploitation.

For example; 90% of the world's fish harvest comes from within national exclusive economic zones, most of which are within the sight of shore. This type of practice has led to a problem that has cumulative effects. The addition of other activities adds to the strain placed on this environment. As a whole, human activity in the coastal zone generally degrades the systems by taking unsustainable quantities of resources. The effects are further exacerbated with the input of pollutant wastes. This provides the need for management. Due to the complex nature of human activity in this zone a holistic approach is required to obtain a sustainable outcome.

The Concept of Sustainability

The concept behind the idea of ICZM is sustainability. For ICZM to succeed, it must be sustainable. Sustainability entails a continuous process of decision making, so there is never an end-state just a readjustment of the equilibrium between development and the protection of the environment. The concept of Sustainability or sustainable development came to fruition in the 1987 report of the World Commission on Environment and Development, Our Common Future. It stated sustainable development is "to meet the needs of the present without compromising the ability of future generations to meet their own needs".

Highlighted are three main standpoints which summarise the idea of Sustainable development, they are:

- Economic development to improve the quality of life of people
- Environmentally appropriate development
- Equitable development

To simplify these points, sustainability should acknowledge the right of humans to live a life that is healthy and productive. It should allow for equal distribution of benefits to all people and in doing so protect the environment through appropriate use.

Sustainability is by no means a set of prescriptive actions, more accurately it is a way of thinking. Adapting this way of thinking paves the way for a longer-term view with a more holistic approach, something successful ICZM can achieve.

Expectations of ICZM

As previously stated, for ICZM to be successful it must adhere to the principles that define sustainability and act upon them in ways that are integrated. An optimal balance between environmental protection and the development of economic and social sectors is paramount.

As part of the holistic approach ICZM applies, many aspects within a coastal zone are expected to be considered and accounted for. These include but are not limited to: the spatial, functional, legal, policy, knowledge, and participation dimensions. Below are four identified goals of ICZM:

- Maintaining the functional integrity of the coastal resource systems;
- Reducing resource-use conflicts;
- Maintaining the health of the environment;
- Facilitating the progress of multisectoral development

Failure to include these aspects and goals would lead to a form of unsustainable management, undermining the paradigms explicit to ICZM.

Defining Integration

The term 'integration' can be adopted for many different purposes, it is therefore quite important to define the term in the context of the management of the coastal zone to appreciate the intentions of ICZM. Integration within ICZM occurs in and between many different levels, 5 types of integration that occur within ICZM, are explained below;

Integration Among Sectors: Within the coastal environment there are many sectors that operate. These human activities are largely economic activities such as tourism, fisheries, and port companies. A sense of co-operation between sectors is the main requirement for sector integration within ICZM. This comes from the realisation of a common goal focused around sustainability and the appreciation of one another within the area.

Integration between Land and Water Elements of the Coastal Zone: This is the realisation of the physical environment being a whole. The coastal environment is a dynamic relationship between many processes all of which are interdependent. The link must be made between imposing a change on one system or feature and its inevitable 'flow on' effects.

Integration Among Levels of Government: Between levels of governance, consistency and co-operation is needed throughout planning and policy making. ICZM is most effective where initiatives have common purpose at local, regional, and national levels. Common goals and actions increase efficiency and mitigate confusion.

Integration between Nations: This sees ICZM as an important tool on a global scale. If goals and beliefs are common on a supranational scale, large scale problems could be mitigated or avoided.

Integration Among Disciplines: Throughout ICZM, knowledge should be accepted from all disciplines. All means of scientific, cultural, traditional, political and local expertise need to be accounted for. By including all these elements a truly holistic approach towards management can be achieved.

The term integration in a coastal management context has many horizontal and vertical aspects, which reflects the complexity of the task and it proves a challenge to implement.

ICZM Framework

Management must embrace a holistic viewpoint of the functions that makeup the complex and dynamic nature of interactions in the coastal environment. Management framework must be applied to a defined geographical limit (often complicated) and should operate with a high level of integration. Due to the diverse nature of the world's coastline and coastal environments, it is not possible to create a framework that is 'one-size-fits-all.' Different activities, interests and issues also complicate matters. So management will always be unique to countries, regions and ultimately on a local scale.

A common thought process and decision making framework however, can be fairly uniform as a part of ICZM around the world. To achieve the principles set out in sustainable types of management a step by step process can be adhered to.

Firstly, issues and problems need to be identified and assessments of these need to be quantified. This first step will include integration between government, sectoral entities and local residents. The assessments also have to be broad in their application. Once the issues and problems have been identified and weighted, an effective management plan can be made. The plan will be specific to the area in question. Thirdly, the adoption of the plan can be carried out. They can be legally binding statutory plans, strategies or objectives which are generally quite powerful or they can be non-statutory processes and can act as a guide for future development. This duality is largely beneficial as the future can be taken into account, but still provide for a firm stance based in the present. The fourth step is implementation, this active phase includes; law enforcement, education, development etc.

The implementation activities will be of course, be as unique as their environments and can take many forms. The last phase is evaluation of the whole process. The principles of sustainability mean that there is no 'end state.' ICZM is an ongoing process which should constantly readjust the equilibrium between economic development and the protection of the environment. Feedback is a crucial part of the process and allows for continued effectiveness even when a situation may change.

ICZM in the Mediterranean

At the Conference of the Plenipotentiaries on the ICZM Protocol that took place on 20–21 January 2008 in Madrid, the ICZM Protocol was signed. Under the presidency of the Minister of Environment of Spain, H.E. Ms. Cristina Narbona Ruiz, fourteen Contracting Parties of the Barcelona Convention signed the Protocol. These are the following: Algeria, Croatia, France, Greece, Israel, Italy, Malta, Monaco, Montenegro, Morocco, Slovenia, Spain, Syria and Tunisia. All other Parties announced to do so in the very near future.

This is the 7th Protocol in the framework of the Barcelona Convention, and the decision to approve the draft text and recommendation to the Conference of the Plenipotentiaries to sign it was taken at the 15th Ordinary Meeting of the Contracting Parties during their meeting in Almeria, on 15–18 January 2008. All the parties are convinced that this Protocol is a crucial milestone in the history of MAP(UNEP/MAP). It will allow the countries to better manage their coastal zones, as well as to deal with the emerging coastal environmental challenges, such as the climate change.

The ICZM Protocol is a unique legal instrument in the entire international community and the Mediterranean countries are proud of this fact. They are willing to share these experiences with other coastal countries of the world. The signing of the Protocol came after six years of dedicated work of all the Parties.

In September 2012, Croatia and Morocco ratified the Protocol, which brings the number of ratifications to 9 (Slovenia, Montenegro, Albania, Spain, France, European Union, Syria, Croatia, Morocco).

A road-map for the implementation of the ICZM Process, prepared by the Priority Action Programme (PAP/RAC) is available on the Coastal Wiki platform of the PEGASO and ENCORA projects: ICZM Process.

Constraints of ICZM

Major constraints of ICZM are mostly institutional, rather than technological. The 'top-down' approach of administrative decision making sees problematisation as a tool promoting ICZM through the idea of sustainability. Community-based 'bottom-up' approaches can perceive problems and issues that are specific to a local area. The benefit of this is that the problems are real and acknowledged rather than searched for to fit an imposed strategy or policy. Public consultation and involvement is very important for current 'top-down' approaches, as it can incorporate this 'bottom-up' idea into the policies made. Prescriptive 'top-down' methods have not able to effectively address problems of resource utilization in poor coastal communities as perceptions of the coastal zone differ with regard to developed and developing countries. This leads on to another constraint to ICZM, the idea of common property.

The coastal environment has huge historical and cultural connections with human activity. Its wealth of resources have provided for millennia, with regard to ICZM how does management become legally binding if the dominant perception of the coast is of a common area available to all? And should it? Enforcing restrictions or change to activities within the coastal zone can be difficult as these resources are often very important to people's livelihoods. The idea of the coast being common property fouls 'top-down' approaches. The idea of common property itself is not all that clean, This perception can lead to cumulative exploitation of resources – the very problem this management seeks to extinguish.

ICZM: The New Zealand Case Study

New Zealand is quite unique as it uses sustainable management within legislation, with a high level of importance placed on to the coastal environment. The Resource Management Act (RMA) (1991) promoted sustainable development and mandated the preparation of a New Zealand Coastal Policy Statement (NZCPS), a national framework for coastal planning. It is the only national policy statement that was mandatory. All subsequent planning must not be inconsistent with the NZCPS, making it a very important document. Regional authorities are required to produce Regional coastal policy plans under the RMA (1991) but strangely enough, they only need to include the marine environment seaward of the mean high water mark. But many regional councils have chosen to integrate the 'dry' landward area within their plans, breaking down the artificial barriers. This attempt at ICZM is still in its early days running into many legislative hurdles and is yet to achieve a fully ecosystems-based approach. But as part of ICZM, evaluation and adoption of changes is important and ongoing changes to the NZCPS in the form of reviews is currently happening. This will provide an excellent stepping stone for future initiatives and the development of a fully integrated form of coastal management.

ICZM: The Iran Case Study

Preparation of comprehensive management plans for optimum utilization of existent sources and potentials in all developed and developing countries is one of the appropriate approaches for constant and permanent utilization of natural, human and financial sources. The versatility of natural sources in coastal areas has made private and governmental users and investors to participate in this section to gain the utmost profits. Therefore, the necessity of preparation and implementation of management plans for perpetual utilization of existent sources in coastal areas has become inevitable. Iran, possessing some 6000 km of coastline in north and south, owns abundant economic capacities in coastal zones and regarding the versatility of nature and coast operators and management of coastal activities and operations, necessity of attention to Integrated Coastal Zone

Management becomes more significant. Such necessity has gained its legal support through ratification of arrangements no. 40 from transportation chapter of third and article no. 63 of fourth economic, social and cultural economic schedule and its executive regulations. The General Director of coasts and ports engineering of Ports and Maritime Organization was detailed to take the studies of ICZM into consideration. The first phase of these studies began in spring 2003 and was fulfilled in autumn 2006.

The outcome of this phase was compilation of following reports accomplished by several national and international skilled consultants: 1- Project Methodology 2- Scrutinized scope of services related to studies 3- Investigation of studies' needs and project preparation and performance 4- Study, definition and determination of Iranian coastal zones boundaries 6- Investigation of International concepts, methods and experiences about Integrated Coastal Zone Management 7- Study and investigation of different features of Integrated Coastal Zone Management in Iran 8- Preparation and designation of geographic database 9- Purchasing and preparing basic data The second phase of studies started up in autumn 2005 and since then this phase has been fully accomplished and presented, In which six competent Iranian consultants with some cooperation of international consultants are responsible for preparing the eleven results of second part of the studies.

The European Union

The European Parliament and the European Council "adopted in 2002 a Recommendation on Integrated Coastal Zone Management which defines the principles sound coastal planning and management. These include the need to base planning on sound and shared knowledge, the need to take a long-term and cross-sector perspective, to pro-actively involve stakeholders and the need to take into account both the terrestrial and the marine components of the coastal zone".

Conclusion

The Integrated Coastal Zone Management (ICZM) appears to be a key element for the sustainable development of these zones. However this recent notion may not be adapted to all cases. The natural disasters Sumatra earthquake and the Indian Ocean tsunami have made a lot of impact on the coastal environment and also the stakeholder's perception on mitigation and management of coastal hazards. Successful implementation is still a major challenge to the idea of ICZM .

Coastal Development Hazards

A coastal development hazard is something that affects the natural environment by man-made products.

As coasts become more developed, the vulnerability component of the equation increases as there is more value at risk to the hazard. The likelihood component of the equation also increases in terms of there being more value on the coast so a higher chance of hazardous situation occurring. Fundamentally humans create hazards with their presence. In a coastal example, erosion is a process that happens naturally on the Canterbury Bight as a part of the coastal geomorphology of the area and strong long shore currents. This process becomes a hazard when humans interact with that coastal environment by developing it and creating value in that area.

In Burton 1978 *'The Environment as Hazard'* a natural hazard is defined as the release of energy or materials that threaten humans or what they value. In a coastal context these hazards vary temporally and spatially from a rare, sudden, massive release of energy and materials such as a major storm event or tsunami, to the continual chronic release of energy and materials such long-term coastal erosion or sea-level rise. It is this type coastal hazard, specifically around erosion and attributes surrounding erosion that this article will focus on.

Coastal Population Growth and Development on Coasts

Globally the number of people living on the coast is increasing. It has been stated that there has been over a 35% increase in the population of people living on the coasts since 1995. The average density of people in coastal regions is 3 times higher than the global average density. Historically city development especially large cities were based on coasts due to the economic benefits of the ports. In 1950 there were only 2 megacities (cities with greater than 8 million people) in the coastal zone, London and New York. By the mid-nineties there were 13. Although coastal areas globally have shown population growth and increases in density, very few in-depth quantitative global studies of population have been carried out, especially in terms of distribution across specific environs, like coasts. The spatial distribution and accuracy of global data must be significantly improved before realistic quantitative assessments of the global impacts of coastal hazards can be made. As currently much of the data is collected and analysed in the aftermath of disasters.

Historic studies have put estimates of the number of deaths due to cyclones over the last 200 years around the Bay of Bengal exceeding 1.3 million. However in developed countries, as can be expected, the death toll is significantly lower but the economic losses due to coastal hazards are increasing. The United States of America (USA) for example had major losses in Hurricane Andrew, which hit Florida and Louisiana in 1992.

This rushing to the coast is exhibited in property value. A study by Bourassa et al. (2004) found in Auckland New Zealand, wide sea views contributed on average an additional 59% to the value of a waterfront property. This effect diminished rapidly the further from the property was from the coast. In another study it was found that moving 150m away from the Gulf of Mexico lowered property values by 36%.

Insurance premiums in coastal hazard areas are an inconsequential determinant of property values, given the significant amenity values provided by the coast in terms of views and local recreation. Sea-level rise, coastal erosion, and the exacerbating interaction between these two natural phenomena are likely to pose a significant threat for the loss of capital assets in coastal areas in the future. It is hard to say if the vulnerability to coastal hazards by those residing there is perceived yet dominated by the amenity value of coasts or simply ignored.

Coastal Erosion Hazards

Coastal erosion is one of the most significant hazards associated with the coast. Not in terms of a rare massive release of energy or material resulting in loss of life, as is associated with tsunami and cyclones, but in terms of a continual chronic release that forms a threat to infrastructure, capital assets and property.

Beach Erosion Process

Storm induced large erosion events are a part of the natural evolutionary process of fine sediment, gently sloping beaches. Increased wave energy in storms leads to the removal of foreshore, berm and dune sediments. These displaced sediments are then deposited as near shore sand bars and act to dampen the wave energy lessoning the amount of sediment that is being eroded from the coast. When wave energies decrease post storm events, the sediments from these newly deposited near shore bars are returned to the upper beach, rebuilding the berm. This self-correcting cycle is an active balance between wave energies and fine sediment deposition. This store of sediment being available for erosion in storms and re-depositing when the event has subsided is an important natural buffer mechanism against protecting the mainland from erosion and minimising coastal retreat.

Dune Destruction

Sand dunes are very dynamic fragile structures that act as stores of sediment used to carry out the coastal processes mentioned above. This removal of the upper beach sediments is important from a hazard perspective as this is the area of the coast that is often utilised for property development due to the high prices sea front properties with a view can achieve. In Pegasus Bay, New Zealand, storm events in 1978 and 2001 caused significant erosion of the New Brighton and Waimairi sand beaches. In the 1978 storm event houses on the seaward side of the New Brighton Spit suffered from undercutting as the dune sediment in which they were built on was eroded by high wave energy.

This same storm event caused similar erosion damage to houses built on the upper dunes in Raumati, on the west coast of the North Island, New Zealand. Bulldozing and bulk removal of sand from protective coastal dunes is therefore an extremely hazardous activity, and one that has been widely carried out in New Zealand in order to form a surface on which to build on to obtain sea views.

Canterbury Bight

In Kirk (2001) coastal erosion on the Canterbury bight, South Canterbury was said to have reached up to 8 m per year. This coastal process could be measured in more ways than one, the afore mentioned distance of coastal retreat or the decreased dollar value of developed assets, land and infrastructure that are at risk.

To date, erosion on the Canterbury Bight has led to the loss of agricultural land, threatened valuable infrastructure including holiday settlements, and reduced coastal lagoons and wetlands.

Historically, erosion on the Canterbury Bight was a natural process, but has now been exacerbated by human intervention. The Waitaki River was the dominant source of sediments for the beaches between Oamaru and Timaru. Since the damming of the Waitaki River in 1935 erosion of the coastal cliffs has become the primary source of sediment in the north flowing current moving up the coast of South Canterbury.

Erosion Mitigation

Engineered Structures: This destruction of sand dunes is often then mitigated with construction of seawalls, revetments and groynes in often futile attempts to prevent storm erosion hazards to unsuitably located assets and infrastructure on coasts.

These engineered methods are commonly ineffective and frequently actually magnify the hazard or just move the hazard down coast. In Porthcawl, South Wales, a seawall constructed to stop erosion in 1887 was replaced in 1906, 1934 and finally in 1984 when the beach was paved as each prior structure was undermined by further erosion.

The loss of aesthetics due to the lack of a sand beach resulted in tourists utilising alternative beaches. Therefore incurring an even greater economic loss on top of the cost of the engineering.

Restorative Dune Planting

The alternative to hard engineering measures is sand dune conservation. This involves protecting the sand dunes and allowing the natural buffering processes to occur. Dune protection and conservation can be facilitated in a number of ways, actively with dune planting and sand fencing, or with better planning by developing away from or well behind the dune structures not on them. On the New Brighton Spit the spread of marram grass *(Ammophila arenaria)* has resulted in effective dune stabilisation in areas. However this invasive exotic species has mostly replaced indigenous species like pingao *(Desmoschoenus spiralis)* meaning that although the stability of the coastal area has gained, the historic, native cultural values of the area have suffered.

Beach Nourishment

A further soft-engineering method for protecting the shoreline is beach nourishment, due to cost this is a solution that has been used primarily for the benefit of the tourism industry. As a result of erosion Miami Beach had almost no stored sediment left by the mid 1970s, consequently, visitor numbers declined and development of the area decreased. A beach nourishment programme was set up resulting in an influx of development and infrastructure in the late 1970s. Miami Beach was rejuvenated to such an extent that annual revenue from foreign tourists alone is $2.4 billion, compared to the $52 million cost of the 20-year nourishment project. Tax revenue from tourists who visit Miami Beach alone more than covers the cost of beach nourishment projects across the nation. Using the capitalised annual cost of the project, for every $1 that has been invested annually on the nourishment, Miami Beach has received almost $500 annually in foreign exchange.

Coastal Geography

Coastal geography is the study of the dynamic interface between the ocean and the land, incorporating both the physical geography (i.e. coastal geomorphology, geology and oceanography) and the human geography (sociology and history) of the coast. It involves an understanding of coastal weathering processes, particularly wave action, sediment movement and weather, and also the ways in which humans interact with the coast. Coastal geography is that branch of geography, incorporating physical and human geography,which deals with the study of the dynamic interface between ocean and land.

Wave Action and Longshore Drift

Figure: *Port Campbell in southern Australia is a high-energy shoreline.*

The waves of different strengths that constantly hit against the shoreline are the primary movers and shapers of the coastline. Despite the simplicity of this process, the differences between waves and the rocks they hit result in hugely varying shapes.

The effect that waves have depends on their strength. Strong, also called destructive waves occur on high-energy beaches and are typical of Winter. They reduce the quantity of sediment present on the beach by carrying it out to bars under the sea. Constructive, weak waves are typical of low-energy beaches and occur most during summer. They do the opposite to destructive waves and increase the size of the beach by piling sediment up onto the berm.

One of the most important transport mechanisms results from wave refraction. Since waves rarely break onto a shore at right angles, the upward movement of water onto the beach (swash) occurs at an oblique angle. However, the return of water (backwash) is at right angles to the beach, resulting in the net movement of beach material laterally. This movement is known as beach drift. The endless cycle of swash and backwash and resulting beach drift can be observed on all beaches. This may differ between coasts.

Probably the most important effect is longshore drift (LSD)(Also known as Littoral Drift), the process by which sediment is continuously moved along beaches by wave action. LSD occurs because waves hit the shore at an angle, pick up sediment (sand) on the shore and carry it down the beach at an angle (this is called swash). Due to gravity, the water then falls back perpendicular to the beach, dropping its sediment as it loses energy (this is called backwash). The sediment is then picked up by the next wave and pushed slightly further down the beach, resulting in a continual movement of sediment in one direction. This is the reason why long strips of coast are covered in sediment, not just the areas around river mouths, which are the main sources of beach sediment. LSD is reliant on a constant supply of sediment from rivers and if sediment supply is stopped or sediment falls into a submarine canals at any point along a beach, this can lead to bare beaches further along the shore.

LSD helps create many landforms including barriers, bay beaches and spits. In general LSD action serves to straighten the coast because the creation of barriers cuts off bays from the sea while sediment usually builds up in bays because the waves there are weaker (due to wave refraction), while sediment is carried away from the exposed headlands. The lack of sediment on headlands removes the protection of waves from them and makes them more vulnerable to weathering while the gathering of sediment in bays (where longshore drift is unable to remove it) protects the bays from further erosion and makes them pleasant recreational beaches.

Atmospheric Processes

- Onshore winds blowing "up" the beach, pick up sand and move it up the beach to form sand dunes.
- Rain hits the shore and erodes rocks, and carries weathered material to the shoreline to form beaches.
- Warm weather can encourage biological processes to occur more rapidly. In tropical areas some plants and animals protect stones from weathering, while other plants and animals actually eat away at the rocks.
- Temperatures that vary from below to above freezing point result in frost weathering, whereas weather more than a few degrees below freezing point creates sea ice.

Biological Processes

In tropical regions in particular, plants and animals not only affect the weathering of rocks but are a source of sediment themselves. The shells and skeletons of many organisms are of calcium carbonate and when this is broken down it forms sediment, limestone and clay.

Physical Processes

The main physical Weathering process on beaches is salt-crystal growth. Wind carries salt spray onto rocks, where it is absorbed into small pores and cracks within the rocks. There the water evaporates and the salt crystallises, creating pressure and often breaking down the rock. In some beaches calcium carbonate is able to bind together other sediments to form beachrock and in warmer areas dunerock. Wind erosion is also a form of erosion, dust and sand is carried caround in the air and slowly erodes rock, this happens in a similar way in the sea were the salt and sand is washed up onto the rocks.

Sea Level Changes (Eustatic Change)

The sea level on earth regularly rises and falls due to climatic changes. During cold periods more of the Earth's water is stored as ice in glaciers while during warm periods it is released and sea levels rise to cover more land. Sea levels are currently quite high, while just 18,000 years ago during the Pleistocene ice age they were quite low. Global warming may result in further rises in the future, which presents a risk to coastal cities as most would be flooded by only small rises. As sea levels rise, fjords and rias form. Fjords are flooded glacial valleys and rias are flooded river valleys. Fjords typically have steep rocky sides, while rias have dendritic drainage patterns typical of drainage zones. As tectonic plates move about the Earth they can rise and fall due to changing pressures and the presence of glaciers. If a beach is moving upwards relative to other plates this is known as isostatic change and raised beaches can be formed.

Land Level Changes (Isostatic Change)

This is found in the U.K. as above the line from the Wash to the Severn estuary, the land was covered in ice sheets during the last ice age. The weight of the ice caused northeast Scotland to sink, displacing the southeast and forcing it to rise. As the ice sheets receded the reverse process happened, as the land was released from the weight. At current estimates the southeast is sinking at a rate of about 2 mm per year, with northeast Scotland rising by the same amount.

Coastal Landforms

Spits: If the coast suddenly changes direction, especially around an estuary, spits are likely to form. Long shore drift pushes sediment along the beach but when it reaches a turn as in the diagram, the long shore drift does not always easily turn with it, especially near an estuary where the outward flow from a river may push sediment away from the coast. The area may be also be shielded from wave action, preventing much long shore drift. On the side of the headland receiving weaker waves, shingle and other large sediments will build up under the water where waves are not strong enough to move them along. This provides a good place for smaller sediments to build up to sea level. The sediment, after passing the headland will accumulate on the other side and not continue down the beach, sheltered both by the headland and the shingle.

Slowly over time sediment simply builds on this area, extending the spit outwards, forming a barrier of sand. Once in a while, the wind direction will change and come from the other direction. During this period the sediment will be pushed along in the other direction. The spit will start to grow backwards, forming a 'hook'. After this time the spit will grow again in the original direction. Eventually the spit will not be able to grow any further because it is no longer sufficiently sheltered from erosion by waves, or because the estuary current prevents sediment resting. Usually in the salty but calm waters behind the spit there will form a salt marshland. Spits often form around the breakwater of artificial harbours requiring dredging.

Occasionally, if there is no estuary then it is possible for the spit to grow across to the other side of the bay and form what is called a bar, or barrier. Barriers come in several varieties, but all form in a manner similar to spits. They usually enclose a bay to form a lagoon. They can join two headlands or join a headland to the mainland. When an island is joined to the mainland with a bar or barrier it is known as a tombolo. This usually occurs due to wave refraction, but can also be caused by isostatic change, a change in the level of the land (e.g. Chesil Beach). An example of this is along the Holderness coastline.

Coastal Erosion

Coastal erosion is the wearing away of land and the removal of beach or dune sediments by wave action, tidal currents, wave currents or drainage. Waves, generated by storms, wind, or fast moving motor craft, cause coastal erosion, which may take the form of long-term losses of sediment and rocks, or merely the temporary redistribution of coastal sediments; erosion in one location may result in accretion nearby. The study of erosion and sediment redistribution is called 'coastal morphodynamics'. It may be caused by hydraulic action, abrasion, impact and corrosion.

On non-rocky coasts, coastal erosion results in dramatic (or non-dramatic) rock formations in areas where the coastline contains rock layers or fracture zones with varying resistance to erosion. Softer areas become eroded much faster than harder ones, which typically result in landforms such as tunnels, bridges, columns, and pillars. Also abrasion commonly happens in areas where there are strong winds, loose sand,and soft rocks. The blowing of millions of sharp sand grains creates a sandblasting effects. This effect helps to erode, smooth and polish rocks. The definition of abrasion is grinding and wearing away of rock surfaces through the mechanical action of other rock or sand particles.

Examples

An example of a cliffed coast is in north-south Wales where over years of the sea beating at the cliffs the houses on top have begun collapsing into the sea. With some of them you can even see inside as the entire backs of some of the houses have come off and have been launched over the clifftop that has been torn through by the ferocious sea.

Dunwich, the capital of the English medieval wool trade, disappeared over the period of a few centuries due to redistribution of sediment by waves. Human interference can also increase coastal erosion: Hallsands in Devon, England, was a coastal village that washed away over the course of a year, an event directly caused by dredging of shingle in the bay in front of it.

The California coast, which has soft cliffs of sedimentary rock and is heavily populated, regularly has incidents of housing damage as cliffs erode. Damage in Pacifica is shown in the gallery below. Devil's Slide, Santa Barbara and Malibu are regularly affected.

The Holderness coastline on the east coast of England, just north of the Humber Estuary, is one of the fastest eroding coastline in Europe due to its soft clay cliffs and powerful waves. Groynes and other artificial measures to keep it under control has only accelerated the process further down the coast, because longshore drift starves the beaches of sand, leaving them more exposed. The White Cliffs of Dover have also been affected.

Wave Action

Hydraulic Action: Hydraulic action occurs when waves striking a cliff face compress air in cracks on the cliff face. This exerts pressure on the surrounding rock, and can progressively splinter and remove pieces. Over time, the cracks can grow, sometimes forming a cave. The splinters fall to the sea bed where they are subjected to further wave action.

Attrition

Attrition occurs when waves causes loose pieces of rock debris (scree) to collide with each other, grinding and chipping each other, progressively becoming smaller, smoother and rounder. Scree also collides with the base of the cliff face, chipping small pieces of rock from the cliff or have a corrasion (abrasion) effect, similar to sandpapering.

Corrasion and Corrosion

Corrasion (abrasion) occurs when waves break on cliff faces and slowly erode it. As the sea pounds cliff faces it also uses the scree from other wave actions to batter and break off pieces of rock from higher up the cliff face which can be used for this same wave action and attrition.

Corrosion or solution/chemical weathering occurs when the sea's pH (anything below pH 7.0) corrodes rocks on a cliff face. Limestone cliff faces, which have a moderately high pH, are particularly affected in this way. Wave action also increases the rate of reaction by removing the reacted material.

Factors that Influence Erosion Rates

Primary Factors: The ability of waves to cause erosion of the cliff face depends on many factors.

The hardness or 'erodibility' of sea-facing rocks is controlled by the rock strength and the presence of fissures, fractures, and beds of non-cohesive materials such as silt and fine sand.

The rate at which cliff fall debris is removed from the foreshore depends on the power of the waves crossing the beach. This energy must reach a critical level to remove material from the debris lobe. Debris lobes can be very persistent and can take many years to completely disappear.

Beaches dissipate wave energy on the foreshore and provide a measure of protection to the adjoining land.

The stability of the foreshore, or its resistance to lowering. Once stable, the foreshore should widen and become more effective at dissipating the wave energy, so that fewer and less powerful waves reach beyond it. The provision of updrift material coming onto the foreshore beneath the cliff helps ensure a stable beach.

The adjacent bathymetry, or configuration of the seafloor, controls the wave energy arriving at the coast, and can have an important influence on the rate of cliff erosion. Shoals and bars offer protection from wave erosion by causing storm waves to break and dissipate their energy before reaching the shore. Given the dynamic nature of the seafloor, changes in the location of shoals and bars may cause the locus of beach or cliff erosion to change position along the shore.

Coastal erosion has been greatly affected by the rising sea levels globally. There has been great measures of increased coastal erosion on the Eastern seaboard of the United States. Locations such as Florida have noticed increased coastal erosion. In reaction to these increases Florida and its individual counties have increased budgets to replenish the eroded sands that attract visitors to Florida and help support its multi-billion dollar tourism industries.

Secondary Factors

- Weathering and transport slope processes
- Slope hydrology
- Vegetation
- Cliff foot erosion
- Cliff foot sediment accumulation
- Resistance of cliff foot sediment to attrition and transport
- Human Activity

Tertiary Factors

- Resource extraction
- Coastal management

Beach Evolution

The shoreline is where the land meets the sea and it is continually changing. Over the long term, the water is eroding the land. Beaches represent a special case, in that they exist where sand accumulated from the same processes that strip away rocky and sedimentary material. I.e., they can grow as well as erode. River deltas are another exception, in that silt that erodes up river can accrete at the river's outlet and extend ocean shorelines. Catastrophic events such as tsunamis, hurricanes and storm surges accelerate beach erosion, potentially carrying away the entire sand load. Human activities can be as catastrophic as hurricanes, albeit usually over a longer time interval.

Erosion and Accretion

Extraordinary Processes: tsunamis and hurricane-driven storm surges:

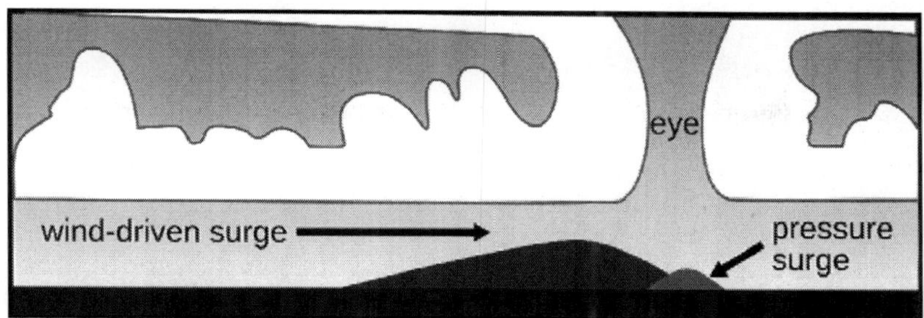

Tsunamis, potentially enormous waves often caused by earthquakes, have great erosional and sediment-reworking potential. They may strip beaches of sand that may have taken years to accumulate and may destroy trees and other coastal vegetation. Tsunamis are also capable of flooding hundreds of metres inland past the typical high-water level and fast-moving water, associated with the inundating tsunami, can crush homes and other coastal structures.

A storm surge is an onshore gush of water associated with a low pressure weather system—storms. Storm surges can cause beach accretion and erosion. Historically notable storm surges occurred during the North Sea Flood of 1953, Hurricane Katrina, and the 1970 Bhola cyclone.

Gradual Processes

The gradual evolution of beaches often comes from the interaction of longshore drift, a wave-driven process by which sediments move along a beach shore, and other sources of erosion or accretion, such as nearby rivers.

Deltas

Deltas are nourished by alluvial systems and accumulate sand and silt, growing where the sediment flux from land is large enough to avoid complete removal by coastal currents, tides, or waves. Most modern deltas formed during the last five thousand years, after the present sea-level high stand was attained. However, not all sediment remains permanently in place: in the short term (decades to centuries), exceptional river floods, storms or other energetic events may remove significant portions of delta sediment or change its lobe distribution and, on longer geological time scales, sea-level fluctuations lead to destruction of deltaic features.

Historical Accretion of European Beaches

In the Mediterranean sea, deltas have been continuously growing for the last several thousand years. Six to seven thousand years ago, the sea level stabilized, and continuous river systems, ephemeral torrents, and other factors began this steady accretion.

Since intense human use of coastal areas is a relatively recent phenomenon (except in the Nile delta), beach contours were primarily shaped by natural forces until the last centuries.

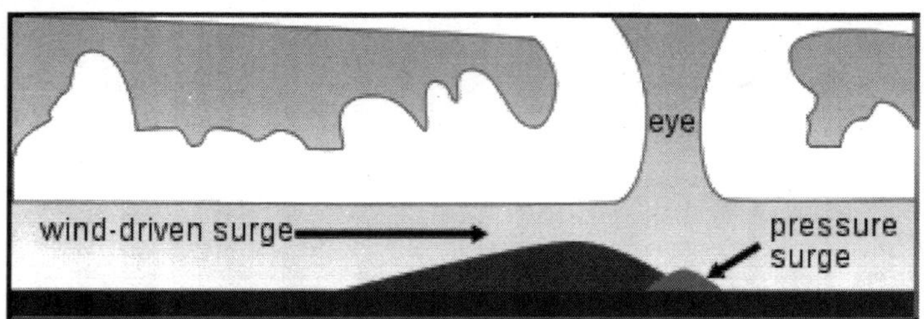

In Barcelona, for example, the accretion of the coast was a natural process until the late Middle Ages, when harbor-building increased the rate of accretion.

The port of Ephesus, one of the great cities of the Ionian Greeks in Asia Minor, was filled with sediment due to accretion from a nearby river; it is now 5 kilometres (3.1 mi) from the sea. Likewise, Ostia, the once-important port near ancient Rome, is now several kilometres inland, the coastline having moved slowly seaward.

Bruges became a port during the early Middle Ages and was accessible by sea until around 1050. At that time, however, the natural link between Bruges and the sea silted up. In 1134, a storm flood opened a deep channel, the Zwin, linking the city to the sea until the fifteenth century via a canal from the Zwin to Bruges. Bruges had to use a number of outports, such as Damme and Sluis, for this purpose. In 1907, a new seaport was inaugurated in Zeebrugge.

Modern Beach Recession

At the present time important segments of low coasts are in recession, losing sand and reducing beach dimensions. This loss can occur very rapidly. Examples of this are occurring at Sète, in California, in Poland, in Aveiro (Portugal), and in Holland and elsewhere along the North Sea. In Europe, coastal erosion is widespread (at least 70%) and distributed very irregularly.

Relative Sea Level Changes

Several geological events and the climate can change (progressively or suddenly) the relative height of the Earth's surface to the sea-level. These events or processes continuously change coastlines.

Volcanism and Earthquakes

Volcanic activity can create new islands. The 800 metres (2,600 ft) in diameter Surtsey Island, Iceland, for example, was created between November 1963 and June 1967. The island has since partially eroded, but it is expected to last another 100 years.

Some earthquakes can create sudden variations of relative ground level and change the coastline dramatically. Structurally controlled coasts include the San Andreas fault zone in California and the seismic Mediterranean belt (from Gibraltar to Greece).

The Bay of Pozzuoli, in Pozzuoli, Italy experienced hundreds of tremors between August 1982 and December 1984. The tremors, which reached a peak on October 4, 1983, damaged 8,000 buildings in the city centre and raised the sea bottom by almost 2 metres (6.6 ft). This rendered the Bay of Pozzuoli too shallow for large craft and required the reconstruction of the harbour with new quays. The photo at the upper right shows the harbor before the uplift while the one on the bottom right shows the new quay.

Gradual Processes: Subsidence and Uplift

Subsidence is the motion of the Earth's surface downward relative to the sea level due to internal geodynamic causes. The opposite of subsidence is uplift, which increases elevation.

Venice is probably the best-known example of a subsiding location. It experiences periodic flooding when extreme high tides or surges arrive. This phenomenon is caused by the compaction of young sediments in the Po River delta area, magnified by subsurface water and gas exploitation. Man-made works to solve this progressive sinking have been unsuccessful.

Mälaren, the third-largest lake in Sweden, is an example of deglacial uplift. It was once a bay on which seagoing vessels were once able to sail far into the country's interior, but it ultimately became a lake. Its uplift was caused by deglaciation: the removal of the weight of ice-age glaciers caused rapid uplift of the depressed land. For 2,000 years as the ice was unloaded, uplift proceeded at about 7.5 centimeters (3.0 in)/year. Once deglaciation was complete, uplift slowed to about 2.5 centimeters (0.98 in) annually, and it decreased exponentially after that. Today, annual uplift rates are 1 centimeter (0.39 in) or less, and studies suggest that rebound will continue for about another 10,000 years. The total uplift from the end of deglaciation may be up to 400 metres (1,300 ft).

Longshore Drift

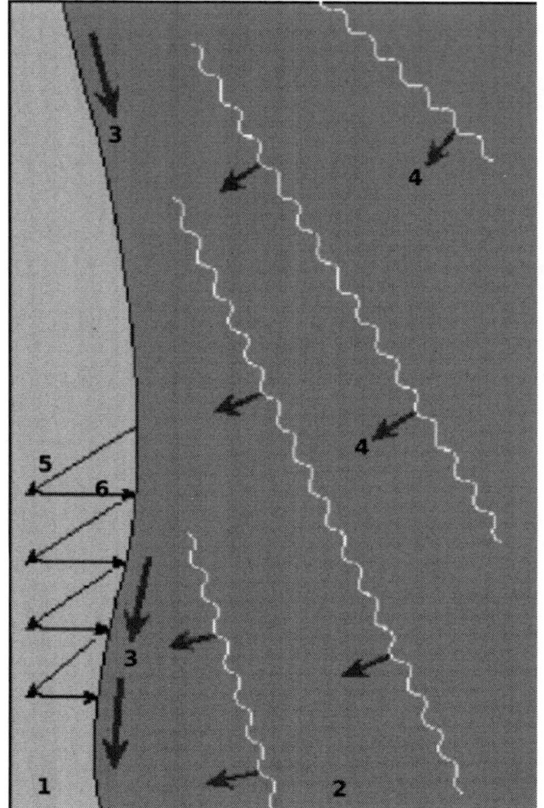

Figure: *Diagram demonstrating longshore drift*

1=beach

2=sea

3=longshore current direction

4=incoming waves

5=swash

6=backwash

Longshore drift consists of the transportation of sediments (clay, silt, sand and shingle) along a coast at an angle to the shoreline, which is dependent on prevailing wind direction, swash and backwash. This process occurs in the littoral zone, and in or close to the surf zone. The process is also known as longshore transport or littoral drift.

Longshore drift is influenced by numerous aspects of the coastal system, with processes that occur within the surf zone largely influencing the deposition and erosion of sediments. Longshore currents can generate oblique breaking waves which result in longshore transport.

Longshore drift can generally be defined in terms of the systems within the surf zone. Sediment transport along the shore and surf zone is influenced by the swash (occurs in the direction of prevailing wind), which moves the pebble up the beach at the angle moves the pebble back down the beach due to the influence of gravity.

Longshore drift affects numerous sediment sizes as it works in slightly different ways depending on the sediment (e.g. the difference in long shore drift of sediments from a sandy beach to that of sediments from a shingle beach). Sand is largely affected by the oscillatory force of breaking waves, the motion of sediment due to the impact of breaking waves and bed shear from long shore current. Whereas because shingle beaches are much steeper than sandy ones, plunging breakers are more likely to form, causing the majority of long shore transport to occur in the swash zone, due to a lack of surf zone.

Overview

Longshore Drift Formulas: There are numerous calculations that take into consideration the factors that produce longshore drift. These formulations are:

1. Bijker formula (1967,1971)
2. The Engelund and Hansen formula (1967)
3. The Ackers and White formula (1973)
4. The Bailard and Inman formula (1981)
5. The Van Rijn formula (1984)
6. The Watanabe formula (1992)

These formulas all provide a different view into the processes that generate longshore drift. The most common factors taken into consideration in these formulas are:

- Suspended and bed load transport
- Waves e.g. breaking and non-breaking
- The shear exerted by waves or the flow associated with waves.

Features of Shoreline Change

Longshore drift plays a large role in the evolution of a shoreline, as if there is a slight change of sediment supply, wind direction, or any other coastal influence longshore drift can change dramatically, impacting on the formation and evolution of a beach system or profile.

These changes do not occur due to one factor within the coastal system, in fact there are numerous alterations that can occur within the coastal system that may affect the distribution and impact of longshore drift. Some of these are:

1. Geological changes, e.g. erosion, backshore changes and emergence of headlands.
2. Change in hydrodynamic forces, e.g. change in wave diffraction in headland and offshore bank environments.
3. Change to hydrodynamic influences, e.g. the influence of new tidal inlets and deltas on drift.
4. Alterations of the sediment budget, e.g. switch of shorelines from drift to swash alignment, exhaustion of sediment sources.
5. The intervention of humans, e.g. cliff protection, groynes, detached breakwaters.

The Sediment Budget

The sediment budget takes into consideration sediment sources and sinks within a system. This sediment can come from any source with examples of sources and sinks consisting of:

- Rivers
- Lagoons
- Eroding land sources
- Artificial sources e.g. nourishment
- Artificial sinks e.g. mining/extraction
- Offshore transport
- Deposition of sediment on shore

This sediment then enters the coastal system and is transported by longshore drift. A good example of the sediment budget and longshore drift working together in the coastal system is inlet ebb-tidal shoals, which store sand that has been transported by long shore transport. As well as storing sand these systems may also transfer or by pass sand into other beach systems, therefore inlet ebb-tidal shoal systems provide a good sources and sinks for the sediment budget.

Sediment deposition (geology) throughout a shoreline profile conforms to the null point hypothesis; where gravitational and hydraulic forces determine the settling velocity of grains in a seaward fining sediment distribution. Long shore occurs in a 90 degree backwash so it would be presented as a right angle with the wave line.

Natural Features

Spits

Spits are formed when longshore drift travels past a point (e.g. river mouth or re-entrant) where the dominant drift direction and shoreline do not veer in the same direction. As well as dominant drift direction, spits are affected by the strength of wave driven current, wave angle and the height of incoming waves.

Spits are landforms that have two important features, with the first feature being the region at the up-drift end or proximal end (Hart et al., 2008). The proximal end is constantly attached to land (unless breached) and may form a slight "barrier" between the sea and an estuary or lagoon. The second important spit feature is the down-drift end or distal end, which is detached from land and in some cases, may take a complex hook-shape or curve, due to the influence of varying wave directions.

As an example, the New Brighton spit in Canterbury, New Zealand, was created by longshore drift of sediment from the Waimakariri River to the north. This spit system is currently in equilibrium but undergoes phases of deposition and erosion.

Barriers

Barrier systems are attached to the land at both the proximal and distal end and are generally widest at the down-drift end. These barrier systems may enclose an estuary or lagoon system, like that of Lake Ellesmere enclosed by the Kaitorete Spit or hapua which form at river-coast interface such as at the mouth of the Rakaia River.

The Kaitorete Spit in Canterbury, New Zealand, is a barrier/spit system (which generally falls under the definition barrier, as both ends of the landform are attached to land, but has been named a spit) that has existed below Banks Peninsula for the last 8000 years. This system has undergone numerous changes and fluctuations due to avulsion of the Waimakariri River (which now flows to the north or Banks Peninsula), erosion and phases of open marine conditions. The system underwent further changes c.500 year BP, when longshore drift from the eastern end of the "spit" system created the barrier, which has been retained due to ongoing longshore transport.

Tidal Inlets

The majority of tidal inlets on longshore drift shores accumulate sediment in flood and ebb shoals. Ebb-deltas may become stunted on highly exposed shores and in smaller spaces, whereas flood deltas are likely to increase in size when space is available in a bay or lagoon system. Tidal inlets can act as sinks and sources for large amounts of material, which therefore impacts on adjacent parts of the coastline.

The structuring of tidal inlets is also important for longshore drift as if an inlet is unstructured sediment may by pass the inlet and form bars at the down-drift part of the coast. Although this may also depend on the inlet size, delta morphology, sediment rate and by passing mechanism. Channel location variance and amount may also influence the impact of long shore drift on a tidal inlet as well. For example, the Arcachon lagoon is a tidal inlet system in South west France, which provides large sources and sinks for longshore drift sediments. The impact of longshore drift sediments on this inlet system is highly influenced by the variation in the number of lagoon entrances and the location of these entrances. Any change in these factors can cause severe down-drift erosion or down-drift accretion of large swash bars.

Human Influences

Groynes

Groynes are shore protection structures, placed at equal intervals along the coastline in order to stop coastal erosion and generally cross the intertidal zone. Due to this, groyne structures are usually used on shores with low net and high annual longshore drift in order to retain the sediments lost in storm surges and further down the coast.

There are numerous variations to groyne designs with the three most common designs consisting of:

1. zig-zag groynes, which dissipate the destructive flows that form in wave induced currents or in breaking waves.
2. T-head groynes, which reduce wave height through wave diffraction.
3. 'Y' head, a fish tail groyne system.

Artificial Headlands

Artificial headlands are also shore protection structures, which are created in order to provide a certain amount of protection to beaches or bays. Although the creation of headlands involves accretion of sediments on the up-drift side of the headland and moderate erosion of the down-drift end of the headland, this is undertaken in order to design a stabilised system that allows material to accumulate in beaches further along the shore. Artificial headlands can occur due to natural accumulation or also through artificial nourishment.

Detached Breakwaters

Detached breakwaters are shore protection structures, created to build up sandy material in order to accommodate drawdown in storm conditions. In order to accommodate drawdown in storm conditions detached breakwaters have no connection to the shoreline, which lets currents and sediment pass between the breakwater and the shore. This then forms a region of reduced wave energy, which encourages the deposition of sand on the lee side of the structure. Detached breakwaters are generally used in the same way as groynes, to build up the volume of material between the coast and the breakwater structure in order to accommodate storm surges.

Ports and Harbours

The creation of ports and harbours throughout the world can seriously impact on the natural course of longshore drift. Not only do ports and harbours pose a threat to longshore drift in the short term, they also pose a threat to shoreline evolution. The major influence the creation of a port or harbour can have on longshore drift is the alteration of sedimentation patterns, which in turn may lead to accretion and/or erosion of a beach or coastal system.

As an example, the creation of a port in Timaru, New Zealand in the late 1800s led to a significant change in the longshore drift along the South Canterbury coastline. Instead of longshore drift transporting sediment north up the coast towards the Waimataitai lagoon, the creation of the port blocked the drift of these (coarse) sediments and instead caused them to accrete to the south of the port at South beach in Timaru. The accretion of this sediment to the south, therefore meant a lack of sediment being deposited on the coast near the Waimataitai lagoon (to the north of the port), which led to the loss of the barrier enclosing the lagoon in the 1930s and then shortly after, the loss of the lagoon itself. As with the Waimataitai lagoon the Washdyke Lagoon, which currently lies to the north of the Timaru port is undergoing erosion and may eventually breach causing loss of another lagoon environment.

Marine Terrace

A marine terrace, coastal terrace, raised beach or perched coastline is a relatively flat, horizontal or gently inclined surface of marine origin, mostly an old abrasion platform which has been lifted out of the sphere of wave activity (sometimes called "tread"). Thus it lies above or under the current sea level, depending on its time of formation. It is bounded by a steeper ascending slope on the landward side and a steeper descending slope on the seaward side (sometimes called "riser"). Due to its reasonably flat shape it is often used for anthropogenic structures like settlements and infrastructure.

Morphology

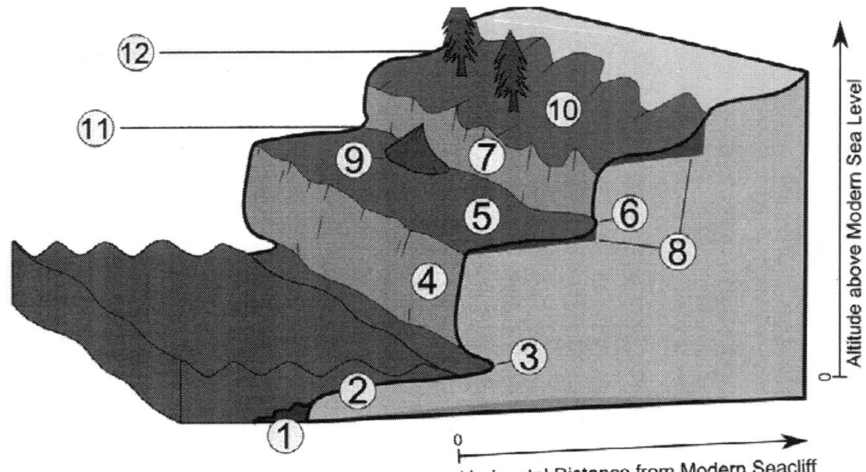

Figure: *Typical sequence of erosional marine terraces. 1) low tide cliff/ramp with deposition, 2) modern shore (wave-cut/abrasion-) platform, 3) notch/inner edge, modern shoreline angle, 4) modern sea cliff, 5) old shore (wave-cut/abrasion-) platform, 6) paleo-shoreline angle, 7) paleo-sea cliff, 8) terrace cover deposits/marine deposits, colluvium, 9) alluvial fan, 10) decayed and covered sea cliff and shore platform, 11) paleo-sea level I, 12) paleo-sea level II. - after various authors*

The platform of a marine terrace usually has a gradient between 1°- 5° depending on the former tidal range with commonly a linear to concave profile. The width is very variable, reaching up to 1000 m, and seems to differ between northern and southern hemisphere. The cliff faces, delimiting the platform, can vary in steepness depending on the relative roles of marine and subaerial processes.

At the intersection of the former shore (wave-cut/abrasion-) platform and the rising cliff face it commonly retains a shoreline angle or inner edge (notch) which indicates the location of the shoreline at the time of maximum sea ingression and therefore a paleo sea level. Sub-horizontal platforms usually terminate in a low tide cliff and it is believed that the occurrence of these platforms depends on tidal activity. Marine terraces can extend for several tens of kilometers parallel to the coast.

Older terraces are covered by marine and/or alluvial or colluvial materials while the uppermost terrace levels usually are less well-preserved. While marine terraces in areas of relatively rapid uplift rates (> 1 mm/year) can often be correlated to individual interglacial periods or stages, those in areas of slower uplift rates may have a polycyclic origin with stages of returning sea levels after times of exposure to weathering.

Marine terraces can be covered by a wide variety of soils with complex histories and different ages. In protected areas allochtonous sandy parent materials from tsunami deposits may be found. Common soil types found on marine terraces include planosols and solonetz.

Formation

Causes: The formation of marine terraces is controlled by changes in environmental conditions and by tectonic activity during recent geological times. Changes in climatic conditions have led to eustatic sea-level oscillations and isostatic movements of the Earth's crust, especially with the changes between glacial and interglacial periods.

Processes of eustasy lead to glacioeustatic sea level fluctuations due to changes of the water volume in the oceans and hence to regressions and transgressions of the shoreline. At times of maximum glacial extent during the last glacial period the sea level was about 100 m lower compared to today.

Eustatic sea level changes can also be caused by changes of the void volume of the oceans, either through sedimento-eustasy or tectono-eustasy. Processes of isostasy involve the uplift of continental crusts including their shorelines. Today the process of glacial isostatic adjustment mainly applies to Pleistocene glaciated areas. In Scandinavia, for instance, the present rate of uplift reaches up to 10 mm/year.

In general, eustatic marine terraces were formed during separate sea level highstands of interglacial stages and can be correlated to marine oxygene isotopic stages (MIS). Glacioisostatic marine terraces were mainly created during stillstands of the isostatic uplift. When eustasy was the main factor for the formation of marine terraces, derived sea level fluctuations can indicate former climate changes. This conclusion has to be treated with care, as isostatic adjustments and tectonic activities can be extensively overcompensated by a eustatic sea level rise. Thus, in areas of both eustatic and isostatic or tectonic influences, the course of the relative sea level curve can be complicated. Hence most of today's marine terrace sequences were formed by a combination of tectonic coastal uplift and Quaternary sea level fluctuations.

Jerky tectonic uplifts can also lead to marked terrace steps while smooth relative sea level changes may not result in obvious terraces and their formations are often not referred to as marine terraces.

Processes

Marine terraces often result from marine erosion along rocky coast lines in temperate regions due to wave attack and sediment carried in the waves. Erosion also takes place in connection with weathering and cavitation.

The speed of erosion is highly dependent on the shoreline material (hardness of rock), the bathymetry and the bedrock properties and can be between only a few millimeters per year for granitic rocks and more than 10 m per year for volcanic ejecta. The retreat of the sea cliff generates a shore (wave-cut/abrasion-) platform through the process of abrasion. A relative change of the sea level leads to regressions or transgressions and eventually forms another terrace (marine-cut terrace) at a different altitude while notches in the cliff face indicate short stillstands.

It is believed that the terrace gradient increases with tidal range and decreases with rock resistance. In addition the relationship between terrace width and the strength of the rock is inverse and higher rates of uplift and subsidence as well as a higher slope of the hinterland increases the number of terraces formed during a certain time.

Furthermore shore platforms are formed by denudation and marine-built terraces arise from accumulations of materials removed by shore erosion. Thus a marine terrace can be formed by both erosion and accumulation. However, there is an ongoing debate about the roles of wave erosion and weathering in the formation of shore platforms. Reef flats or uplifted coral reefs are another kind of marine terrace found in intertropical regions. They are a result of biological activity, shoreline advance and accumulation of reef materials.

While a terrace sequence can date back hundreds of thousands of years, its degradation is a rather fast process. On the one hand a deeper transgression of cliffs into the shoreline may completely destroy previous terraces; on the other hand older terraces might be decayed or covered by deposits, colluvia or alluvial fans. Erosion and backwearing of slopes caused by incisive streams play another important role in this degradation process.

Mapping and Surveying

For exact interpretations of the morphology extensive datings, surveying and mapping of marine terraces is applied. This includes stereoscopic aerial photographic interpretation (ca. 1 : 10,000 - 25,000), on-site inspections with topographic maps (ca. 1 : 10,000) and analysis of eroded and accumulated material. Moreover the exact altitude can be determined with an aneroid barometer or preferably with a levelling instrument mounted on a tripod. It should be measured with the accuracy of 1 cm and at about every 50 – 100 m, depending on the topography. In remote areas technics of photogrammetry and tacheometry can be applied.

Correlation and Dating

Different methods for dating and correlation of marine terraces can be used and combined.

Correlational Dating

The morphostratigraphic approach focuses especially in regions of marine regression on the altitude as the most important criterion to distinguish coast lines of different ages. Moreover individual marine terraces can be correlated based on their size and continuity. Also paleo-soils as well as glacial, fluvial, eolian and periglacial landforms and sediments may be used to find correlations between terraces. On New Zealand's North Island, for instance, tephra and loess were used to date and correlate marine terraces. At the terminus advance of former glaciers marine terraces can be correlated by their size, as their width decreases with age due to the slowly thawing glaciers along the coast line.

The lithostratigraphic approach uses typical sequences of sediment and rock strata to prove sea level fluctuations on the basis of an alternation of terrestrial and marine sediments or littoral and shallow marine sediments. Those strata show typical layers of transgressive and regressive patterns. However, an unconformity in the sediment sequence might make this analysis difficult.

The biostratigraphic approach uses remains of organisms which can indicate the age of a marine terrace. For that often mollusc shells, foraminifera or pollen are used. Especially Mollusca can show specific properties depending on their depth of sedimentation. Thus they can be used to estimate former water depths. Marine terraces are often correlated to marine oxygene isotopic stages (MIS) (e.g. Johnson, M. E.; Libbey, L. K. 1997) and can also be roughly dated using their stratigraphic position.

Direct Dating

There are various methods for the direct dating of marine terraces and their related materials including ^{14}C radiocarbon dating, which is the most common one. E.g. this method has been used on the North Island of New Zealand to date several marine terraces. It utilizes terrestrial biogenic materials in coastal sediments such as mollusc shells analyzing the ^{14}C isotope.

In some cases dating based on the $^{320}Th/^{234}U$ ratio was applied though in case of detrital contamination or low uranium concentrations a high resolution dating was found to be difficult. In a study in southern Italy paleomagnetism was used to carry out paleomagnetic datings and luminescence dating (OSL) was used in different studies on the San Andreas Fault and on the Quaternary Eupcheon Fault in South Korea.

In the last decennia, the dating of marine terraces has been enhanced since the arrival of terrestrial cosmogenic nuclides method, and particularly through the use of ^{10}Be and ^{26}Al cosmogenic isotopes produced in-situ. These isotopes record the duration of surface exposure to cosmic rays. and this exposure age reflects the age of abandonment of a marine terrace by the sea.

Relevance for Other Research Areas

Marine terraces play an important role in the research on tectonics and earthquakes. They may show patterns and rates of tectonic uplift and thus may be used to estimate the tectonic activity in a certain region. In some cases the exposed secondary landforms can be correlated with known seismic events such as the 1855 Wairarapa earthquake on the Wairarapa Fault near Wellington, New Zealand which produced a 2.7 m uplift. This can be estimated from the vertical offset between raised shorelines in the area.

Furthermore with the knowledge of eustatic sea level fluctuations the speed of isostatic uplift can be estimated and eventually the change of relative sea levels for certain regions can be reconstructed. Thus marine terraces also provide information for the research on climate change and trends in future sea level changes.

When analyzing the morphology of marine terraces it must be considered, that both eustasy and isostasy can have an influence on the formation process. This way can be assessed, whether there were changes in sea level or whether tectonic activities took place.

Prominent Examples

Marine terraces can be found on many geodynamically influenced coastlines around the world. Important sites include various coasts of New Zealand, e.g. Turakirae Head near Wellington being one of the world's best and most thoroughly studied examples. Also along the Cook Strait in New Zealand there is a well-defined sequence of uplifted marine terraces from the late Quaternary at Tongue Point.

It features a well preserved lower terrace from the last interglacial, a widely eroded higher terrace from the penultimate interglacial and another still higher terrace, which is nearly completely decayed. Furthermore on New Zealand's North Island at the eastern Bay of Plenty a sequence of seven marine terraces has been studied.

Along many coasts of mainland and islands around the Pacific, marine terraces are typical coastal features. An especially prominent marine terraced coastline can be found north of Santa Cruz, near Davenport, California, where terraces probably have been raised by repeated slip earthquakes on the San Andreas Fault. Hans Jenny (pedologist) famously researched the pygmy forests of the Mendocino and Sonoma county marine terraces. The marine terrace's "ecological staircase" of Salt Point State Park is also bound by the San Andreas Fault.

Along the coasts of South America marine terraces are present, where the highest ones are situated where plate margins lie above subducted oceanic ridges and the highest and most rapid rates of uplift occur.

At Cape Laundi, Sumba Island, Indonesia an ancient patch reef can be found at 475 m above sea level as part of a sequence of coral reef terraces with eleven terraces being wider than 100 m. The coral marine terraces at Huon Peninsula, New Guinea, which extend over 80 km and rise over 600 m above present sea level are currently on UNESCO's tentative list for world heritage sites under the name *Houn Terraces - Stairway to the Past.*

Other considerable examples include marine terraces rising up to 360 m on some Philippine Islands and along the Mediterranean Coast of North Africa, especially in Tunisia, rising up to 400 m.

Raised Shoreline

A raised shoreline is an ancient shoreline exposed above current water level. These landforms are formed by a relative change in sea level due to global sea level rise, isostatic rebound, and/or tectonic uplift.

These surfaces are usually exposed above modern sea level when a heavily glaciated area experiences a glacial retreat, causing water levels to rise. This area will then experience post-glacial rebound, effectively raising the shoreline surface.

Examples of raised shorelines can be found along the coasts of formerly glaciated areas in Ireland and Scotland, as well as in North America. Raised shorelines are exposed at various locations around the Puget Sound of Washington State.

Wave-cut Platform

A wave-cut platform, coastal benches, wave-cut benches or shore platform is the narrow flat area often found at the base of a sea cliff or along the shoreline of a lake, bay, or sea that was created by the action of waves. Wave-cut platforms are often most obvious at low tide when they become visible as huge areas of flat rock. Sometimes the landward side of the platform is covered by sand, forming the beach, and then the platform can only be identified at low tides or when storms move the sand.

Formation

It forms after destructive waves hit against the cliff face, causing undercutting between the high and low water marks, mainly as a result of corrosion and hydraulic power, creating a wave-cut notch. This notch then enlarges into a cave.

The waves undermine this portion until the roof of the cave cannot hold due to the pressure and freeze-thaw weathering acting on it, and collapses, resulting in the cliff retreating landward.

The base of the cave forms the wave-cut platform as attrition causes the collapsed material to be broken down into smaller pieces, while some cliff material may be washed into the sea. This may be deposited at the end of the platform, forming an off-shore terrace.

Because of the continual wave action, a wave-cut platform represents an extremely hostile environment and only the toughest of organisms can utilize such a niche.

Use of Ancient Examples

Ancient wave-cut platforms provide evidence of past sea and lake levels. Raised and abandoned platforms, sometimes found behind modern beaches, are evidence of higher sea levels in the geological past, and have been used to identify areas of isostatic adjustment. By using scientific dating methods, or examination of marine fossils found on the platform, it is possible to work out when the platform was formed, thus giving geographers and geologists information about sea levels at known times in the past. This has been used in the United Kingdom and other previously glaciated areas to calculate the rate at which land is rising now that it is no longer covered in ice.

Where the coastline itself is changing due to seismic action, there may be a series of platforms showing earlier sea levels and indicating the amount of uplift caused by various earthquakes.

Usage of Term 'Wave-cut'

According to Trenhaile, Sunamura, and Massalink and Hughes, the term 'wave-cut platform' should no longer be used as it assumes that shore platforms are the result of wave action, which is not always true. Shore platforms, like comparable river and lake platforms, are erosional features that develop when removal of saprock and other debris by waves and currents leaves behind a bedrock surface below the water table.

Submersion (Coastal Management)

Submersion is the sustainable cyclic portion of coastal erosion where coastal sediments move from the visible portion of a beach to the submerged nearshore region, and later return to the original visible portion of the beach. The recovery portion of the sustainable cycle of sediment behaviour is (accretion).

Submersion vs Erosion

The sediment that is submerged during rough weather forms landforms including storm bars. In calmer weather waves return sediment to the visible part of the beach. Due to longshore drift some sediment can end up further along the beach from where it started. Often coastal areas have developed sustainable coastal positions where the sediment moving off beaches is sustainable submersion. Unfortunately, for many inhabited coastlines, anthropogenic interference in coastal processes has meant that erosion is often more permanent than submersion.

Community Perception

The term *erosion* often is associated with undesirable impacts on the environment, whereas submersion should be celebrated as a sustainable part of healthy foreshores. Communities making decisions about coastal management need to develop understanding of the components of beach recession and be able to separate the component that is temporary sustainable submersion from the more serious irreversible anthropogenic or climate change erosion portion.

Artificial Reef (Costal Management)

An artificial reef is a human-made underwater structure, typically built to promote marine life in areas with a generally featureless bottom, control erosion, block ship passage, or improve surfing. Many reefs are built using objects that were built for other purposes, for example by sinking oil rigs (through the Rigs-to-Reefs programme), scuttling ships, or by deploying rubble or construction debris. Other artificial reefs are purpose built (e.g. the reef balls) from PVC or concrete. Shipwrecks may become artificial reefs when preserved on the sea floor. Regardless of construction method, artificial reefs generally provide hard surfaces where algae and invertebrates such as barnacles, corals, and oysters attach; the accumulation of attached marine life in turn provides intricate structure and food for assemblages of fish.

History

The construction of artificial reefs is thousands of years old. Ancient Persians blocked the mouth of the Tigris River to thwart Indian pirates by building an artificial reef, and during the First Punic War the Romans built a reef across the mouth of the Carthaginian harbor in Sicily to trap the enemy ships within and assist in driving the Carthaginians from the island. Artificial reefs to increase fish yields or for algaculture have been used at least since 17th century Japan, when rubble and rocks were used to grow kelp, while the earliest recorded construction of artificial reef in the United States is from the 1830s when logs from huts were used off the coast of South Carolina to improve fishing.

Since at least the 1830s, American fishermen used interlaced logs to build artificial reefs. More recently, castaway junk, such as old refrigerators, shopping carts, ditched cars, out-of-service vending machines replaced the logs in ad hoc reefs. Officially sanctioned projects have incorporated decommissioned subway cars, vintage battle tanks, armored personnel carriers and oil drilling rigs.

Development

Artificial reefs tend to develop in more or less predictable stages. First, where an ocean current encounters a vertical structure, it can create a plankton-rich upwelling that provides a reliable feeding spot for small fish such as sardines and minnows, which draw in pelagic predators like bluefin tuna and sharks.

Next come creatures seeking protection from the ocean's lethal openness—hole and crevice dwellers such as grouper, snapper, squirrelfish, eels, and triggerfish.

Opportunistic predators such as jack and barracuda also appear, waiting for their prey to venture out. Over months and years the reef structure becomes encrusted with algae, tunicates, hard and soft corals, and sponges.

Reefs

Florida: Florida is the site of many artificial reefs, many created from deliberately sunken ships, including Coast Guard cutters Duane and Bibb and the U.S. Navy landing ship Spiegel Grove.

Osborne Reef

In the early 1970s, more than 2,000,000 used vehicle tires were dumped off the coast of Fort Lauderdale, Florida to form an artificial reef. Catastrophically, the tires were not properly secured to the reef structures.

Ocean currents broke them loose, sending them crashing into the developing reef and its natural neighbours. As of 2009, less than 100,000 of the tires had been removed after more than 10 years of efforts.

Neptune Reef: Neptune Reef was originally conceived as an art project that would gradually decay. Burial at sea became a way of financing the project. As of 2011, about 200 "placements" had taken place.

Cremated remains are mixed with cement and either encased in columns or molded into sea-star, brain-coral, 15 foot-tall castings of lions or other shapes before entering the water.

The ex-USS Oriskany

The world's largest artificial reef was created by the purposeful sinking of the aircraft carrier USS *Oriskany* off the coast of Pensacola, Florida, in 2006.

Figure: *Sea life growing on the remains of the USS Oriskany, intentionally sunk in 2006 to become an artificial reef.*

The ex-*USNS* Hoyt S. Vandenberg

The second-largest artificial reef is the USNS *Hoyt S. Vandenberg*, a former World War II era troop transport that served as a spacecraft tracking ship after the war. The Vandenberg was scuttled seven miles off Key West on May 27, 2009, in 140 feet of clear water. Supporters expect the ship to draw recreational divers away from natural reefs, allowing those reefs to recover from damage from overuse.

The ex-*USS* Spiegel Grove

The ex-Spiegel Grove is located on Dixie Shoal, 6 miles (10 km) off the Florida Keys in the Florida Keys National Marine Sanctuary. Her exact location is 25°042 00.23 N 80°182 00.73 W.

Delaware

Redbird Reef: In late 2000, The MTA New York City Transit decided to phase out its outdated fleet of subway cars to make room for the R142 and R142A trains. These subway cars, (Redbirds), ran on the IRT lines in the New York City Subway system for 40 years. Each car was sold, stripped, decontaminated, loaded on a barge, and sunk in the Atlantic Ocean. Some had number plates removed because of rust, and auctioned off on eBay. 1,200 subway cars were sunk.

In September 2007 the MTA approved a contract worth $6 million, to send 1,600 of its retired subway cars to be used as artificial reefs. Most of these trains ran on The BMT/IND lines. The trains include the R32, R38, R40 and R42. These models are stainless steel. The MTA will replace them with the R160A and R160B trains. The plastic front ends were removed before sinking. The retired fleet included old work trains and cars damaged beyond repair.

Mexico

Cancun Underwater Museum: Since November 2009, artist Jason Decaires Taylor has created more than 400 life size sculptures off the coast of Cancun, Mexico. The coral reefs in this region suffered heavy degradation due to repetitive hurricane abuse. This project funded by The National Marine Park and the Cancun Nautical Association was designed to emmulate coral reefs using a neutral ph clay. Taylor has constructed unique settings depicting daily activities ranging from a man watching TV to a 1970's replica of a Volkswagen Beetle. This artificial reef has relieved pressure from the nearby Manchones Reef. The design and materials implemented in this project have proved the ecological viability of artificial reefs.

Australia

Since the late 1990s, the Australian government has been providing decommissioned warships for use as artificial reefs for recreational scuba diving. So far, the following six ships have been sunk:

- Ex-HMAS Swan at Dunsborough in Western Australia during December 1998.
- Ex-HMAS Perth at Albany in Western Australia during November 2001.
- Ex-HMAS Hobart in Yankalilla Bay in South Australia during November 2002.
- Ex-HMAS Brisbane off the Sunshine Coast in Queensland during July 2005.
- Ex-HMAS Canberra at a site west of the entrance to Port Phillip Bay in Victoria during October 2009.
- Ex-HMAS Adelaide off Terrigal on the New South Wales Central Coast during April 2011

Gibraltar

The Gibraltar Reef was first proposed by Eric Shaw in 1973. Initial experiments with tyres proved unsuccessful as the tyres were simply swept away by currents or buried underneath sand. In 1974, boats from local marinas and the Gibraltar Port Authority were donated. The first two were barges that were sunk in Camp Bay. In 2006, a 65 ton wooden boat, "True Joy" (also referred to as "Noah's Ark") was sunk here as well, followed by the MV New Flame, a mid sized bulk carrier, in 2007.

In 2013, a dropping of more than 70 concrete blocks of one square metre with metal bars, took place. There was a discussion about, because some experts don't consider the blocks of concrete with metal bars a real reef but a system to bother Spanish fishing activities. The blocks with metal devices proved unpopular with Spanish fishermen who have suffered important economic loses and demanded the process to be stopped because it would prevent them trawling the seabed for fish. The metal bars are designed to break the fishermen nets. The dropping lead to a diplomatic conflict between the Kingdom of Spain and United Kingdom.

Artificial Surfing Reefs

Artificial surfing reefs have been created for surfing, coastal protection, habitat enhancement and coastal research. The world's first attempt was made in El Segundo, near Los Angeles, in California. The next attempt was at Mosman Beach, Perth, Western Australia. This reef was constructed of large granite rocks placed in a pyramidal shape to form an appropriate breaking wave form that would suit surfers. An artificial reef constructed of over 400 massive, geotextile bags (each one larger than a bus) filled with sand was constructed in 2000 at Narrowneck on the Gold Coast of Queensland, Australia. This artificial reef had two objectives: stabilizing beach nourishment and improving surfing conditions.

In the United States, in particular, demanding coastal permit requirements and environmental opposition present major obstacles to building surfing reefs. As of February 2006, the only reef built in the U.S. for surfing is southern California's "Pratte's Reef", which failed to create waves. Reefs built to enhance marine habitat face less environmental opposition, in part because they are in deeper water and further offshore. A number of such man-made reefs exist near Florida and Hawaii.

Artificial surfing reefs typically resemble a "submerged breakwater", and proponents suggest benefits beyond surfing conditions. Many coastlines are subject to powerful waves that crash directly onshore. An artificial reef 150-300 yards offshore might create surfing opportunities and, by dissipating wave energy, make swimming safer and reduce coastal erosion.

The USS *Spiegel Grove* was sunk in 2002 to make an artificial reef.

Europe's first artificial reef was approved in 2008. Construction began August 30, 2008 in Boscombe, Bournemouth, UK, and opened in November 2009.

The multi-purpose reef reef is expected to create waves up to 30% larger and double the number of surfing days annually. Construction on this reef began in June 2008, and was completed in August 2009. Boscombe Reef was built from large sand-filled geotextile containers, totaling 13,000 cubic metres.

Electro Mineral Accretion (EMA)

Mineral accretion involves applying a low voltage current to a metallic structure to cause limestone to crystallize on the surface, to which coral planulae can attach and grow. The electric current also speeds post-attachment growth.

EMA works like charging a battery with a positive pole, the cathode, and a negative pole, the anode. Applying electric current attracts various dissolved minerals to either the cathode or the anode. Chemical reactions take place at both poles. On the anode, bubbles of oxygen and chlorine gas form. These bubbles float to the surface and dissolve into the air. On the cathode, bubbles of hydrogen gas and a limestone precipitate appear.

The voltage is low enough that it can be generated by floating solar panels or from wave motion.

A coalition of scientists named the Global Coral Reef Alliance (GCRA) is developing a technique called the Biorock Process using mineral accretion for reef restoration, mariculture, and shoreline protection.

Environmental Concerns

According to The Ocean Conservancy, a Washington-based environmental group, the Osbourne reef may be an indication that the benefits of artificial reefs need to be re-examined. Jack Sobel, a senior scientist at the group, has said "There's little evidence that artificial reefs have a net benefit," citing concerns such as toxicity, damage to ecosystems and concentrating fish into one place (worsening overfishing).

Accretion (Coastal Management)

Accretion is the process of coastal sediment returning to the visible portion of a beach or foreshore following a submersion event. A sustainable beach or foreshore often goes through a cycle of submersion during rough weather then accretion during calmer periods. If a coastline is not in a healthy sustainable state, then erosion can be more serious and accretion does not fully restore the original volume of the visible beach or foreshore leading to permanent beach loss.

Managed Retreat

In the context of coastal erosion, managed retreat (also managed realignment) allows an area that was not previously exposed to flooding by the sea to become flooded by removing coastal protection. This process is usually in low lying estuarine areas and almost always involves flooding of land that has at some point in the past been claimed from the sea. Managed retreat is often a response to sea level rise exacerbated by local subsidence of the land surface due to isostatic rebound in the north.

Coastal Defence

In the UK the main reason for implementation of Managed Realignment is generally to improve coastal stability, essentially replacing artificial 'hard' coastal defences with natural 'soft' coastal landforms (Pethick 2002). This process can be used to protect areas of land further inland rather than that near the coast by relying on natural defences to absorb or dampen the force of waves.

Habitat Loss

In addition to being used as a means of coastal defence, Managed Realignment has also been used in a number of cases to mitigate for loss of intertidal habitat.

Although land claim has been an important factor for salt marsh loss in the UK in the past (Allen 1992) the majority of current salt marsh loss in the UK is believed to be due to erosion (Morris et al. 2004). This erosion may involve coastal squeeze, where protective sea walls prevent the landward migration of salt marsh in response to sea level rise when sediment supply is limited (Hulme 2005; Morris et al. 2004). Salt marshes are protected under the EU Habitats Directive as well as providing habitat for a number of species protected by the Birds Directive. Following this guidance, the UK's biodiversity action plan aims to prevent net losses to the area of salt marsh present in

1992. It is therefore a legal requirement that all losses in marsh area must be compensated by replacement habitat with equivalent biological characteristics (Crooks et al. 2001). This equates to the need to restore approximately 1.4 km^2 of salt marsh habitat per year in the UK.

Advantages

There are no direct costs apart from that of removing any defences already in place and maintenance costs are very low. Sediment flow is also restored to its natural state, beaches can be naturally replenished due to erosion of the coast, providing protection and the balance of the coastline returns.

Disadvantages

A certain amount of land will inevitably be lost in this process while beaches are being built up resulting in settlements, farmland and other property being destroyed. Because of this, managed retreat is often not a socially acceptable plan and may invoke the need for compensation to land-owners. Intertidal sites are often a rich archaeological resource and the loss of heritage is a factor to be weighed in managed retreat projects.

There are no agreed protocols on the monitoring of MR sites (Atkinson et al. 2001) and, consequently, very few of the sites are being monitored consistently and effectively (Wolters et al. 2005c). Due to the low levels of monitoring there is little evidence on which to base future managed realignment projects. This has led to the results of Managed Realignment schemes being extremely unpredictable.

Examples

In the UK, the first managed retreat site was an area of 8,000 square metres at Northey Island in Essex flooded in 1991, followed by larger sites at Tollesbury and Orplands (1995), Freiston Shore (2001) and Abbott's Hall Farm, at Great Wigborough in the Blackwater Estuary, it is one of the largest managed retreat schemes in Europe. It covers nearly 280 hectares of land on the north side of the estuary (2002) and a number of others. The programme was started by the RSPB - The Royal Society for the Protection of Birds, of whom own Abbott's Hall Farm, they made five breaches in the original old sea wall to allow the held-back sea to flood through to create salt marshland. The marshland over time reverted to its original state before the time of being cultivated, it has become a great site for birds as for the perfect conditions and is a hot-spot breeding ground too.

Current Progress

At present approximately 4 km^2 of salt marsh have been restored by MR in the UK (Mossman et al. In prep). One of the major reasons cited for the slow pace of current salt marsh restoration in the UK (Morris et al. 2004) is the uncertainty associated with the practice (Foresight).

Intertidal Zone

The intertidal zone, also known as the foreshore and seashore and sometimes referred to as the littoral zone, is the area that is above water at low tide and under water at high tide (in other words, the area between tide marks). This area can include many different types of habitats, with many types of animals, such as starfish, sea urchins, and numerous species of coral.

The well-known area also includes steep rocky cliffs, sandy beaches, or wetlands (e.g., vast mudflats). The area can be a narrow strip, as in Pacific islands that have only a narrow tidal range, or can include many metres of shoreline where shallow beach slopes interact with high tidal excursion.

Organisms in the intertidal zone are adapted to an environment of harsh extremes. Water is available regularly with the tides but varies from fresh with rain to highly saline and dry salt with drying between tidal inundations. The action of waves can dislodge residents in the littoral zone. With the intertidal zone's high exposure to the sun, the temperature range can be anything from very hot with full sun to near freezing in colder climates. Some microclimates in the littoral zone are ameliorated by local features and larger plants such as mangroves. Adaptation in the littoral zone allows the use of nutrients supplied in high volume on a regular basis from the sea which is actively moved to the zone by tides. Edges of habitats, in this case land and sea, are themselves often significant ecologies, and the littoral zone is a prime example.

A typical rocky shore can be divided into a spray zone or splash zone (also known as the supratidal zone), which is above the spring high-tide line and is covered by water only during storms, and an intertidal zone, which lies between the high and low tidal extremes. Along most shores, the intertidal zone can be clearly separated into the following subzones: high tide zone, middle tide zone, and low tide zone. The intertidal zone is one of a number of marine biomes or habitats, including estuaries, neritic, surface and deep zones.

Zonation

Marine biologists divide the intertidal region into three zones (low, middle, and high), based on the overall average exposure of the zone. The low intertidal zone, which borders on the shallow subtidal zone, is only exposed to air at the lowest of low tides and is primarily marine in character. The mid intertidal zone is regularly exposed and submerged by average tides. The high intertidal zone is only covered by the highest of the high tides, and spends much of its time as terrestrial habitat. The high intertidal zone borders on the splash zone (the region above the highest still-tide level, but which receives wave splash). On shores exposed to heavy wave action, the intertidal zone will be influenced by waves, as the spray from breaking waves will extend the intertidal zone.

Depending on the substratum and topography of the shore, additional features may be noticed. On rocky shores, tide pools form in depressions that fill with water as the tide rises. Under certain conditions, such as those at Morecambe Bay, quicksand may form.

Low Tide Zone (Lower Littoral)

This subregion is mostly submerged - it is only exposed at the point of low tide and for a longer period of time during extremely low tides. This area is teeming with life; the most notable difference with this subregion to the other three is that there is much more marine vegetation, especially seaweeds. There is also a great biodiversity. Organisms in this zone generally are not well adapted to periods of dryness and temperature extremes. Some of the organisms in this area are abalone, sea anemones, brown seaweed, chitons, crabs, green algae, hydroids, isopods, limpets, mussels, nudibranchs, sculpin, sea cucumber, sea lettuce, sea palms, sea stars, sea urchins, shrimp, snails, sponges, surf grass, tube worms, and whelks. Creatures in this area can grow to larger sizes because there is more available energy in the localized ecosystem. Also, marine vegetation can grow to much greater sizes than in the other three intertidal subregions due to the better water coverage. The water is shallow enough to allow plenty of light to reach the vegetation to allow substantial photosynthetic activity, and the salinity is at almost normal levels. This area is also protected from large predators such as fish because of the wave action and the relatively shallow water.

5

Intertidal Ecology

Intertidal ecology is the study of intertidal ecosystems, where organisms live between the low and high tide lines. At low tide, the intertidal is exposed whereas at high tide, the intertidal is underwater. Intertidal ecologists therefore study the interactions between intertidal organisms and their environment, as well as between different species of intertidal organisms within a particular intertidal community.

The most important environmental and species interactions may vary based on the type of intertidal community being studied, the broadest of classifications being based on substrates - rocky shore and soft bottom communities.

Organisms living in this zone have a highly variable and often hostile environment, and have evolved various adaptations to cope with and even exploit these conditions.

One easily visible feature of intertidal communities is vertical zonation, where the community is divided into distinct vertical bands of specific species going up the shore. Species ability to cope with abiotic factors associated with emersion stress, such as desiccation determines their upper limits, while biotic interactions e.g.competition with other species sets their lower limits.

Intertidal regions are utilized by humans for food and recreation, but anthropogenic actions also have major impacts, with overexploitation, invasive species and climate change being among the problems faced by intertidal communities. In some places Marine Protected Areas have been established to protect these areas and aid in scientific research.

Types of Intertidal Communities

Intertidal habitats can be characterized as having either hard or soft bottoms substrates. Rocky intertidal communities occur on rocky shores, such as headlands, cobble beaches, or human-made jetties. Their degree of exposure may be calculated using the Ballantine Scale. Soft-sediment habitats include sandy beaches, and intertidal wetlands (e.g., mudflats, and salt marshes). These habitats differ in levels of abiotic, or non-living, environmental factors. Rocky shores tend to have higher wave action, requiring adaptations allowing the inhabitants to cling tightly to the rocks. Soft-bottom habitats are generally protected from large waves but tend to have more variable salinity levels. They also offer a third habitable dimension—depth—thus, many soft-sediment inhabitants are adapted for burrowing.

Environment

Because intertidal organisms endure regular periods of immersion and emersion, they essentially live both underwater and on land and must be adapted to a large range of climatic conditions. The intensity of climate stressors varies with relative tide height because organisms living in areas with higher tide heights are emersed for longer periods than those living in areas with lower tide heights. This gradient of climate with tide height leads to patterns of intertidal zonation, with high intertidal species being more adapted to emersion stresses than low intertidal species. These adaptations may be behavioural (i.e. movements or actions), morphological (i.e. characteristics of external body structure), or physiological (i.e. internal functions of cells and organs). In addition, such adaptations generally cost the organism in terms of energy (e.g. to move or to grow certain structures), leading to trade-offs (i.e. spending more energy on deterring predators leaves less energy for other functions like reproduction).

Intertidal organisms, especially those in the high intertidal, must cope with a large range of temperatures. While they are underwater, temperatures may only vary by a few degrees over the year. However, at low tide, temperatures may dip to below freezing or may become scaldingly hot, leading to a temperature range that may approach 30 °C (86 °F) during a period of a few hours. Many mobile organisms, such as snails and crabs, avoid temperature fluctuations by crawling around and searching for food at high tide and hiding in cool, moist refuges (crevices or burrows) at low tide. Besides simply living at lower tide heights, non-motile organisms may be more dependent on coping mechanisms. For example, high intertidal organisms have a stronger stress response, a physiological response of making proteins that help recovery from temperature stress just as the immune response aids in the recovery from infection.

Intertidal organisms are also especially prone to desiccation during periods of emersion. Again, mobile organisms avoid desiccation in the same way as they avoid extreme temperatures: by hunkering down in mild and moist refuges. Many intertidal organisms, including *Littorina* snails, prevent water loss by having waterproof outer surfaces, pulling completely into their shells, and sealing shut their shell opening. Limpets (*Patella*) do not use such a sealing plate but occupy a home-scar to which they seal the lower edge of their flattened conical shell using a grinding action. They return to this home-scar after each grazing excursion, typically just before emersion. On soft rocks, these scars are quite obvious. Still other organisms, such as the algae *Ulva* and *Porphyra*, are able to rehydrate and recover after periods of severe desiccation.

The level of salinity can also be quite variable. Low salinities can be caused by rainwater or river inputs of freshwater. Estuarine species must be especially euryhaline, or able to tolerate a wide range of salinities. High salinities occur in locations with high evaporation rates, such as in salt marshes and high intertidal pools. Shading by plants, especially in the salt marsh, can slow evaporation and thus ameliorate salinity stress. In addition, salt marsh plants tolerate high salinities by several physiological mechanisms, including excreting salt through salt glands and preventing salt uptake into the roots.

In addition to these exposure stresses (temperature, desiccation, and salinity), intertidal organisms experience strong mechanical stresses, especially in locations of high wave action. There are myriad ways in which the organisms prevent dislodgement due to waves. Morphologically, many mollusks (such as limpets and chitons) have low-profile, hydrodynamic shells. Types of substrate attachments include mussels' tethering byssal threads and glues, sea stars' thousands of suctioning tube feet, and isopods' hook-like appendages that help them hold onto intertidal kelps. Higher profile organisms, such as kelps, must also avoid breaking in high flow locations, and they do so with their strength and flexibility. Finally, organisms can also avoid high flow environments, such as by seeking out low flow microhabitats. Additional forms of mechanical stresses include ice and sand scour, as well as dislodgment by water-borne rocks, logs, etc.

For each of these climate stresses, species exist that are adapted to and thrive in the most stressful of locations. For example, the tiny crustacean copepod *Tigriopus* thrives in very salty, high intertidal tidepools, and many filter feeders find more to eat in wavier and higher flow locations. Adapting to such challenging environments gives these species competitive edges in such locations.

Food Web Structure

During tidal immersion, the food supply to intertidal organisms is subsidized by materials carried in seawater, including photosynthesizing phytoplankton and consumer zooplankton. These plankton are eaten by numerous forms of filter feeders—mussels, clams, barnacles, sea squirts, and polychaete worms—which filter seawater in their search for planktonic food sources. The adjacent ocean is also a primary source of nutrients for autotrophs, photosynthesizing producers ranging in size from microscopic algae (e.g. benthic diatoms) to huge kelps and other seaweeds. These intertidal producers are eaten by herbivorous grazers, such as limpets that scrape rocks clean of their diatom layer and kelp crabs that creep along blades of the feather boa kelp *Egregia* eating the tiny leaf-shaped bladelets. Crabs are eaten by Goliath Grouper, which are then eaten by sharks. Higher up the food web, predatory consumers—especially voracious starfish—eat other grazers (e.g. snails) and filter feeders (e.g. mussels). Finally, scavengers, including crabs and sand fleas, eat dead organic material, including dead producers and consumers.

Species Interactions

In addition to being shaped by aspects of climate, intertidal habitats—especially intertidal zonation patterns—are strongly influenced by species interactions, such as predation, competition, facilitation, and indirect interactions. Ultimately, these interactions feed into the food web structure, described above. Intertidal habitats have been a model system for many classic ecological studies, including those introduced below, because the resident communities are particularly amenable to experimentation.

One dogma of intertidal ecology—supported by such classic studies—is that species' lower tide height limits are set by species interactions whereas their upper limits are set by climate variables. Classic studies by Robert Paine established that when sea star predators are removed, mussel beds extend to lower tide heights, smothering resident seaweeds. Thus, mussels' lower limits are set by sea star predation. Conversely, in the presence of sea stars, mussels' lower limits occur at a tide height at which sea stars are unable to tolerate climate conditions.

Competition, especially for space, is another dominant interaction structuring intertidal communities. Space competition is especially fierce in rocky intertidal habitats, where habitable space is limited compared to soft-sediment habitats in which three-dimensional space is available. Mussels are competitively dominant when they are not kept in check by sea star predation. Joseph Connell's research on two types of high intertidal barnacles, *Balanus balanoides*, now *Semibalanus balanoides*, and a *Chthamalus stellatus*, showed that zonation patterns could also be set by competition between closely related organisms. In this example, *Balanus* outcompetes *Chthamalus* at lower tide heights but is unable to survive at higher tide heights. Thus, *Balanus* conforms to the intertidal ecology dogma introduced above: its lower tide height limit is set by a predatory snail and its higher tide height limit is set by climate. Similarly, *Chthamalus*, which occurs in a refuge from competition (similar to the temperature refuges discussed above), has a lower tide height limit set by competition with *Balanus* and a higher tide height limit is set by climate.

Although intertidal ecology has traditionally focused on these negative interactions (predation and competition), there is emerging evidence that positive interactions are also important. *Facilitation* refers to one organism helping another without harming itself. For example, salt marsh plant species of *Juncus* and *Iva* are unable to tolerate the high soil salinities when evaporation rates are high, thus they depend on neighbouring plants to shade the sediment, slow evaporation, and help maintain tolerable salinity levels. In similar examples, many intertidal organisms provide physical structures that are used as refuges by other organisms. Mussels, although they are tough competitors with certain species, are also good facilitators as mussel beds provide a three-dimensional habitat to species of snails, worms, and crustaceans.

All of the examples given so far are of direct interactions: Species A eat Species B or Species B eats Species C. Also important are indirect interactions where, using the previous example, Species A eats so much of Species B that predation on Species C decreases and Species C increases in number. Thus, Species A indirectly benefits Species C. Pathways of indirect interactions can include all other forms of species interactions. To follow the sea star-mussel relationship, sea stars have an indirect negative effect on the diverse community that lives in the mussel bed because, by preying on mussels and decreasing mussel bed structure, those species that are facilitated by mussels are left homeless. Additional important species interactions include mutualism, which is seen in symbioses between sea anemones and their internal symbiotic algae, and parasitism, which is prevalent but is only beginning to be appreciated for its effects on community structure.

Current Topics

Humans are highly dependent on intertidal habitats for food and raw materials, and over 50% of humans live within 100 km of the coast. Therefore, intertidal habitats are greatly influenced by human impacts to both ocean and land habitats. Some of the conservation issues associated with intertidal habitats and at the head of the agendas of managers and intertidal ecologists are:

1. Climate change: Intertidal species are challenged by several of the effects of global climate change, including increased temperatures, sea level rise, and increased storminess. Ultimately, it has been predicted that the distributions and numbers of species will shift depending on their abilities to adapt (quickly!) to these new environmental conditions. Due to the global scale of this issue, scientists are mainly working to understand and predict possible changes to intertidal habitats.
2. Invasive species: Invasive species are especially prevalent in intertidal areas with high volumes of shipping traffic, such as large estuaries, because of the transport of non-native species in ballast water. San Francisco Bay, in which an invasive *Spartina* cordgrass from the east coast is currently transforming mudflat communities into *Spartina* meadows, is among the most invaded estuaries in the world. Conservation efforts are focused on trying to eradicate some species (like *Spartina*) in their non-native habitats as well as preventing further species introductions (e.g. by controlling methods of ballast water uptake and release).
3. Marine protected areas: Many intertidal areas are lightly to heavily exploited by humans for food gathering (e.g. clam digging in soft-sediment habitats and snail, mussel, and algal collecting in rocky intertidal habitats). In some locations, marine protected areas have been established where no collecting is permitted. The benefits of protected areas may spill over to positively impact adjacent unprotected areas. For example, a

greater number of larger egg capsules of the edible snail *Concholepus* in protected vs. non-protected areas in Chile indicates that these protected areas may help replenish snail stocks in areas open to harvesting. The degree to which collecting is regulated by law differs with the species and habitat.

Marine Debris

Marine debris, also known as marine litter, is human-created waste that has deliberately or accidentally been released in a lake, sea, ocean or waterway. Floating oceanic debris tends to accumulate at the centre of gyres and on coastlines, frequently washing aground, when it is known as *beach litter* or tidewrack. Deliberate disposal of wastes at sea is called *ocean dumping*. Naturally occurring debris, such as driftwood, are also present.

With the increasing use of plastic, human influence has become an issue as many types of plastics do not biodegrade. Waterborne plastic poses a serious threat to fish, seabirds, marine reptiles, and marine mammals, as well as to boats and coasts. Dumping, container spillages, litter washed into storm drains and waterways and wind-blown landfill waste all contribute to this problem.

Types of Debris

Researchers classify debris as either land- or ocean-based; in 1991, the United Nations Joint Group of Experts on the Scientific Aspects of Marine Pollution estimated that up to 80% of the pollution was land-based. A wide variety of anthropogenic artifacts can become marine debris; plastic bags, balloons, buoys, rope, medical waste, glass bottles and plastic bottles, cigarette lighters, beverage cans, polystyrene, lost fishing line and nets, and various wastes from cruise ships and oil rigs are among the items commonly found to have washed ashore. Six pack rings, in particular, are considered emblematic of the problem.

Eighty percent of marine debris is plastic. Plastics accumulate because they typically don't biodegrade as many other substances do. They photodegrade on exposure to sunlight, although they do so only under dry conditions, as water inhibits photolysis.

Ghost Nets

Fishing nets left or lost in the ocean by fishermen – ghost nets – can entangle fish, dolphins, sea turtles, sharks, dugongs, crocodiles, seabirds, crabs and other creatures. These nets restrict movement, causing starvation, laceration and infection, and, in animals that breathe air, suffocation.

Nurdles and Plastic Bags

Figure: *A handful of nurdles, spilt from a train in Pineville, Louisiana*

Nurdles, also known as "mermaids' tears", are plastic pellets, typically under five millimetres in diameter, that are a major component of marine debris. They are a raw material in plastics manufacturing, and enter the natural environment when spilled. Weathering produces ever smaller pieces. Nurdles strongly resemble fish eggs.

Plastic waste has reached all the world's oceans. This pollution harms an estimated 100,000 sea turtles and marine mammals and 1,000,000 sea creatures each year. Pelagic plastic pieces in the centre of our ocean's gyres outnumber live marine plankton, and are passed up the food chain to reach all marine life.

Plastic shopping bags can clog digestive tracts when consumed and can cause starvation through restricting the movement of food, or by filling the stomach and tricking the animal into thinking

it is full. A 1994 study of the seabed using trawl nets in the North-Western Mediterranean around the coasts of Spain, France and Italy reported mean concentrations of debris of 1,935 items per square kilometre. Plastic debris accounted for 77%, of which 93% was plastic bags.

Deep-sea Debris

Litter, made from materials that are denser than surface water (such as glasses, metals and some plastics), has been found to spread over the floor of seas and open oceans, where it can become entangled in corals and interfere with other sea-floor life, or even become buried under sediment, making clean-up extremely difficult, especially due to the wide area of its dispersal compared to shipwrecks. Research performed by MBARI found items including plastic bags below 2000m depth off the west coast of North America and around Hawaii.

Sources of Debris

An estimated 10,000 containers at sea each year are lost by container ships, usually during storms. One famous spillage occurred in the Pacific Ocean in 1992, when thousands of rubber ducks and other toys went overboard during a storm. The toys have since been found all over the world, providing a better understanding of ocean currents. Similar incidents have happened before, such as when Hansa Carrier dropped 21 containers (with one notably containing buoyant Nike shoes). In 2007, MSC Napoli beached in the English Channel, dropping hundreds of containers, most of which washed up on the Jurassic Coast, a World Heritage Site.

In Halifax Harbour, Nova Scotia 52% of items were generated by recreational use of an urban park, 14% from sewage disposal and only 7% from shipping and fishing activities. Around four fifths of oceanic debris is from rubbish blown onto the water from landfills, and urban runoff. In the 1987 Syringe Tide, medical waste washed ashore in New Jersey after having been blown from Fresh Kills Landfill. On the remote sub-Antarctic island of South Georgia, fishing-related debris, approximately 80% plastics, are responsible for the entanglement of large numbers of Antarctic fur seals.

Marine litter is even found on the floor of the Arctic ocean.

Great Pacific Garbage Patch

Once waterborne, debris becomes mobile. Flotsam can be blown by the wind, or follow the flow of ocean currents, often ending up in the middle of oceanic gyres where currents are weakest. The Great Pacific Garbage Patch is one such example of this, comprising a vast region of the North Pacific Ocean rich with anthropogenic wastes. Estimated to be double the size of Texas, the area contains more than 3 million tons of plastic. The gyre contains approximately six pounds of plastic for every pound of plankton per cubic metre of seawater.

The oceans may contain as much as one hundred million tons of plastic.

Islands situated within gyres frequently have coastlines flooded by waste that washes ashore; prime examples are Midway and Hawaii. Clean-up teams around the world patrol beaches to attack this environmental threat.

Environmental Impact

Many animals that live on or in the sea consume flotsam by mistake, as it often looks similar to their natural prey. Bulky plastic debris may become permanently lodged in the digestive tracts of these animals, blocking the passage of food and causing death through starvation or infection. Tiny floating plastic particles also resemble zooplankton, which can lead filter feeders to consume them and cause them to enter the ocean food chain. In samples taken from the North Pacific Gyre in 1999 by the Algalita Marine Research Foundation, the mass of plastic exceeded that of zooplankton by a factor of six.

Toxic additives used in plastic manufacturing can leach into their surroundings when exposed to water. Waterborne hydrophobic pollutants collect and magnify on the surface of plastic debris, thus making plastic more deadly in the ocean than it would be on land. Hydrophobic contaminants bioaccumulate in fatty tissues, biomagnifying up the food chain and pressuring apex predators and humans. Some plastic additives disrupt the endocrine system when consumed; others can suppress the immune system or decrease reproductive rates.

Not all anthropogenic artifacts in the oceans are harmful. Iron and concrete typically do little damage to the environment as they generally sink to the bottom and become immobile, and can even provide scaffolding for artificial reefs. Ships and subway cars have been deliberately sunk for that purpose. Some organisms have adapted to live on floating plastic debris, allowing them to disperse all over the world and become invasive species in remote ecosystems.

Debris Removal

Techniques for collecting and removing marine (or riverine) debris include the use of specialized mechanical equipment designed to clean up floating debris such as a Trash Hunter. These debris skimmer boats easily retrieve a wide variety of manmade and organic floating debris from the waterway. Such activities are often undertaken regularly where floating debris poses danger to navigation. For example, the US Army Corps of Engineers reports removing 90 tons of "drifting material" from San Francisco Bay every month. The Corps has been doing this work since 1942, when a seaplane carrying Admiral Chester W. Nimitz collided with a piece of floating debris and sank, costing the life of its pilot. Once debris becomes "beach litter", collection by hand and specialized beach-cleaning machines are used to gather the debris.

Elsewhere, "trash traps" are installed on small rivers to capture waterborne debris before it reaches the sea. For example, South Australia's Adelaide operates a number of such traps, known as "trash racks" or "gross pollutant traps" on the Torrens River, which flows (during the wet season) into Gulf St Vincent.

Laws and Treaties

Ocean dumping is controlled by international law, including:

- The London Convention (1972) – a United Nations agreement to control ocean dumping
- MARPOL 73/78 – a convention designed to minimize pollution of the seas, including dumping, oil and exhaust pollution

European Law

In 1972 and 1974, conventions were held in Oslo and Paris respectively, and resulted in the passing of the OSPAR Convention, an international treaty controlling marine pollution in the north-east Atlantic Ocean. The Barcelona Convention protects the Mediterranean Sea. The Water Framework Directive of 2000 is a European Union directive committing EU member states to free inland and coastal waters from human influence. In the United Kingdom, the Marine and Coastal Access Act is designed to "ensure clean healthy, safe, productive and biologically diverse oceans and seas, by putting in place better systems for delivering sustainable development of marine and coastal environment".

United States Law

In 1972, the United States Congress passed the Ocean Dumping Act, giving the Environmental Protection Agency power to monitor and regulate the dumping of sewage sludge, industrial waste, radioactive waste and biohazardous materials into the nation's territorial waters. The Act was amended sixteen years later to include medical wastes. It is illegal to dispose of any plastic in US waters.

Ownership

Property law, admiralty law and the law of the sea may be of relevance when lost, mislaid, and abandoned property is found at sea.

Salvage law rewards salvors for risking life and property to rescue the property of another from peril. On land the distinction between deliberate and accidental loss led to the concept of a "treasure trove". In the United Kingdom, shipwrecked goods should be reported to a Receiver of Wreck, and if identifiable, they should be returned to their rightful owner.

Activism

A large number of groups and individuals are active in preventing or educating about marine debris. For example, 5 Gyres is an organization aimed at reducing plastics pollution in the oceans, and was one of two organizations that recently researched the Great Pacific Garbage Patch. Heal the Bay is another organization, focusing on protecting California's Santa Monica Bay, by sponsoring Beach Cleanup programmes along with other activities. Marina DeBris is an artist focusing most of her recent work on educating people about beach trash.

Environmental Impact of Shipping

The environmental impact of shipping includes greenhouse gas emissions and oil pollution. Carbon dioxide emissions from shipping is estimated to be 4 to 5 percent of the global total, and estimated by the International Maritime Organization (IMO) to rise by as much as 72 percent by 2020 if no action is taken.

The First Intersessional Meeting of the IMO Working Group on Greenhouse Gas Emissions from Ships took place in Oslo, Norway on 23–27 June 2008. It was tasked with developing the technical basis for the reduction mechanisms that may form part of a future IMO regime to control greenhouse gas emissions from international shipping, and a draft of the actual reduction mechanisms themselves, for further consideration by IMO's Marine Environment Protection Committee (MEPC).

Ballast Water

Ballast water discharges by ships can have a negative impact on the marine environment.

Cruise ships, large tankers, and bulk cargo carriers use a huge amount of ballast water, which is often taken on in the coastal waters in one region after ships discharge wastewater or unload cargo, and discharged at the next port of call, wherever more cargo is loaded. Ballast water discharge typically contains a variety of biological materials, including plants, animals, viruses, and bacteria. These materials often include non-native, nuisance, invasive, exotic species that can cause extensive ecological and economic damage to aquatic ecosystems.

Sound Pollution

Noise pollution caused by shipping and other human enterprises has increased in recent history. The noise produced by ships can travel long distances, and marine species who may rely on sound for their orientation, communication, and feeding, can be harmed by this sound pollution The Convention on the Conservation of Migratory Species has identified ocean noise as a potential threat to marine life.

Ship Impacts

Marine mammals, such a whales and manatees, risk being struck by ships, causing injury and death. For example, if a ship is travelling at a speed of only 15 knots, there is a 79 percent chance of a collision being lethal to a whale.

One notable example of the impact of ship collisions is the endangered North Atlantic right whale, of which 400 or less remain. The greatest danger to the North Atlantic right whale is injury sustained from ship strikes. Between 1970 and 1999, 35.5 percent of recorded deaths were attributed to collisions. During 1999 to 2003, incidents of mortality and serious injury attributed to ship strikes averaged one per year. In 2004 to 2006, that number increased to 2.6. Deaths from collisions has become an extinction threat.

Exhaust Emissions

Exhaust emissions from ships are considered to be a significant source of air pollution, with 18 to 30 percent of all nitrogen oxide and 9 percent of sulphur oxide pollution. "By 2010, up to 40 percent of air pollution over land could come from ships." Sulphur in the air creates acid rain which damages crops and buildings. When inhaled the sulphur is known to cause respiratory problems and even increase the risk of a heart attack. According to Irene Blooming, a spokeswoman for the European environmental coalition Seas at Risk, the fuel used in oil tankers and container ships is high in sulphur and cheaper to buy compared to the fuel used for domestic land use. "A ship lets out around 50 times more sulphur than a lorry per metric tonne of cargo carried." Cities in the U.S. like Long Beach, Los Angeles, Houston, Galveston, and Pittsburgh see some of the heaviest shipping traffic in the nation and have left local officials desperately trying to clean up the air. Increasing trade between the U.S. and China is helping to increase the number of vessels navigating the Pacific and exacerbating many of the environmental problems. To maintain the level of growth China is experiencing, large amounts of grain are being shipped to China by the boat load. The number of voyages are expected to continue increasing.

3.5 to 4 percent of all climate change emissions are caused by shipping. Air pollution from cruise ships is generated by diesel engines that burn high sulphur content fuel oil, also known as bunker oil, producing sulphur dioxide, nitrogen oxide and particulate, in addition to carbon monoxide, carbon dioxide, and hydrocarbons. Diesel exhaust has been classified by EPA as a likely human carcinogen. EPA recognizes that these emissions from marine diesel engines contribute to ozone and carbon monoxide nonattainment (i.e., failure to meet air quality standards), as well as adverse health effects associated with ambient concentrations of particulate matter and visibility, haze, acid deposition, and eutrophication and nitrification of water. EPA estimates that large marine diesel engines accounted for about 1.6 percent of mobile source nitrogen oxide emissions and 2.8 percent of mobile source particulate emissions in the United States in 2000. Contributions of marine diesel engines can be higher on a port-specific basis. Ultra-low sulphur diesel (ULSD) is a standard for defining diesel fuel with substantially lowered sulphur contents. As of 2006, almost all of the petroleum-based diesel fuel available in Europe and North America is of a ULSD type.

As one way to reduce the impact of greenhouse gas emissions from shipping, vetting agency RightShip has developed an online "GHG Emissions Rating" as a systematic way for the industry to compare a ship's CO2 emissions to peer vessels of a similar size and type. Using higher rated ships can deliver significantly lower CO2 emissions across the voyage length.

One source of environmental pressures on maritime vessels recently has come from states and localities, as they assess the contribution of commercial marine vessels to regional air quality problems when ships are docked in port. For instance, large marine diesel engines are believed to contribute 7 percent of mobile source nitrogen oxide emissions in Baton Rouge/New Orleans. Ships can also have a significant impact in areas without large commercial ports: they contribute about 37 percent of total area nitrogen oxide emissions in the Santa Barbara area, and that percentage is expected to increase to 61 percent by 2015. Again, there is little cruise-industry specific data on this issue. They comprise only a small fraction of the world shipping fleet, but cruise ship emissions may exert significant impacts on a local scale in specific coastal areas that are visited repeatedly. Shipboard incinerators also burn large volumes of garbage, plastics, and other waste, producing ash that must be disposed of. Incinerators may release toxic emissions as well.

In 2005 MARPOL Annex VI came into force to combat this problem. As such cruise ships now employ cctv monitoring on the smoke stacks as well as recorded measuring via opacity metre with some also using clean burning gas turbines for electrical loads and propulsion in sensitive areas.

Oil Spills

Most commonly associated with ship pollution are oil spills. While less frequent than the pollution that occurs from daily operations, oil spills have devastating effects. While being toxic to marine life, polycyclic aromatic hydrocarbons (PAHs), the components in crude oil, are very difficult to clean up, and last for years in the sediment and marine environment. Marine species constantly exposed to PAHs can exhibit developmental problems, susceptibility to disease, and abnormal reproductive cycles. One of the more widely known spills was the Exxon Valdez incident in Alaska. The ship ran aground and dumped a massive amount of oil into the ocean in March 1989. Despite efforts of scientists, managers, and volunteers over 400,000 seabirds, about 1,000 sea otters, and immense numbers of fish were killed.

International Regulation

Some of the major international efforts in the form of treaties are the Marine Pollution Treaty, Honolulu, which deals with regulating marine pollution from ships, and the UN Convention on Law of the Sea, which deals with marine species and pollution. While plenty of local and international regulations have been introduced throughout maritime history, much of the current regulations are considered inadequate. "In general, the treaties tend to emphasize the technical features of safety and pollution control measures without going to the root causes of sub-standard shipping, the absence of incentives for compliance and the lack of enforceability of measures." Cruise ships, for example, are exempt from regulation under the US discharge permit system (NPDES, under the Clean Water Act) that requires compliance with technology-based standards. In the Caribbean, many ports lack proper waste disposal facilities, and many ships dump their waste at sea.

Sewage

The cruise line industry dumps 255,000 US gallons (970 m^3) of greywater and 30,000 US gallons (110 m^3) of black water into the sea every day. Blackwater is sewage, wastewater from toilets and medical facilities, which can contain harmful bacteria, pathogens, viruses, intestinal parasites, and harmful nutrients. Discharges of untreated or inadequately treated sewage can cause bacterial and viral contamination of fisheries and shellfish beds, producing risks to public health. Nutrients in sewage, such as nitrogen and phosphorus, promote excessive algal blooms, which consumes oxygen in the water and can lead to fish kills and destruction of other aquatic life. A large cruise ship (3,000 passengers and crew) generates an estimated 55,000 to 110,000 litres per day of blackwater waste.

Due to the environmental impact of shipping, and sewage in particular marpol annex IV was brought into force September 2003 strictly limiting untreated waste discharge. Modern cruise ships are most commonly installed with a membrane bioreactor type treatment plant for all blackwater and greywater, such as Zenon or Rochem which produce near drinkable quality effluent to be re-used in the machinery spaces as technical water.

Cleaning

Greywater is wastewater from the sinks, showers, galleys, laundry, and cleaning activities aboard a ship. It can contain a variety of pollutant substances, including fecal coliforms, detergents, oil and grease, metals, organic compounds, petroleum hydrocarbons, nutrients, food waste, medical and dental waste. Sampling done by the EPA and the state of Alaska found that untreated greywater from cruise ships can contain pollutants at variable strengths and that it can contain levels of fecal coliform bacteria several times greater than is typically found in untreated domestic wastewater. Greywater has potential to cause adverse environmental effects because of concentrations of nutrients and other oxygen-demanding materials, in particular. Greywater is typically the largest source of liquid waste generated by cruise ships (90 to 95 percent of the total). Estimates of greywater range from 110 to 320 litres per day per person, or 330,000 to 960,000 litres per day for a 3,000-person cruise ship.

Solid Waste

Solid waste generated on a ship includes glass, paper, cardboard, aluminium and steel cans, and plastics. It can be either non-hazardous or hazardous in nature. Solid waste that enters the ocean may become marine debris, and can then pose a threat to marine organisms, humans, coastal communities, and industries that utilize marine waters. Cruise ships typically manage solid waste by a combination of source reduction, waste minimisation, and recycling. However, as much as 75 percent of solid waste is incinerated on board, and the ash typically is discharged at sea, although some is landed ashore for disposal or recycling. Marine mammals, fish, sea turtles, and birds can be injured or killed from entanglement with plastics and other solid waste that may be released or disposed off of cruise ships.

On average, each cruise ship passenger generates at least two pounds of non-hazardous solid waste per day. With large cruise ships carrying several thousand passengers, the amount of waste generated in a day can be massive. For a large cruise ship, about 8 tons of solid waste are generated during a one-week cruise. It has been estimated that 24 percent of the solid waste generated by vessels worldwide (by weight) comes from cruise ships. Most cruise ship garbage is treated on board (incinerated, pulped, or ground up) for discharge overboard. When garbage must be off-loaded (for example, because glass and aluminium cannot be incinerated), cruise ships can put a strain on port reception facilities, which are rarely adequate to the task of serving a large passenger vessel.

Bilge Water

On a ship, oil often leaks from engine and machinery spaces or from engine maintenance activities and mixes with water in the bilge, the lowest part of the hull of the ship. Oil, gasoline, and by-products from the biological breakdown of petroleum products can harm fish and wildlife and pose threats to human health if ingested. Oil in even minute concentrations can kill fish or have various sub-lethal chronic effects. Bilge water also may contain solid wastes and pollutants containing high amounts of oxygen-demanding material, oil and other chemicals. A typical large cruise ship will generate an average of 8 metric tons of oily bilge water for each 24 hours of operation.

To maintain ship stability and eliminate potentially hazardous conditions from oil vapours in these areas, the bilge spaces need to be flushed and periodically pumped dry. However, before a bilge can be cleared out and the water discharged, the oil that has been accumulated needs to be extracted from the bilge water, after which the extracted oil can be reused, incinerated, and/or offloaded in port. If a separator, which is normally used to extract the oil, is faulty or is deliberately bypassed, untreated oily bilge water could be discharged directly into the ocean, where it can damage marine life. A number of cruise lines have been charged with environmental violations related to this issue in recent years.

Ocean Acidification

Ocean acidification is the name given to the ongoing decrease in the pH of the Earth's oceans, caused by the uptake of anthropogenic carbon dioxide (CO_2) from the atmosphere. About 30–40% of the carbon dioxide released by humans into the atmosphere dissolves into the oceans, rivers and lakes.

To maintain chemical equilibrium, some of it reacts with the water to form carbonic acid. Some of these extra carbonic acid molecules react with a water molecule to give a bicarbonate ion and a hydronium ion, thus increasing the ocean's "acidity" (H^+ ion concentration). Between 1751 and 1994 surface ocean pH is estimated to have decreased from approximately 8.25 to 8.14, representing an increase of almost 30% in H^+ ion concentration in the world's oceans.

This increasing acidity is thought to have a range of direct undesirable consequences such as depressing metabolic rates in jumbo squid, depressing the immune responses of blue mussels, and coral bleaching.

Other chemical reactions are also triggered which result in an actual net decrease in the amount of carbonate ions available. In the oceans, this makes it more difficult for marine calcifying organisms, such as coral and some plankton, to form biogenic calcium carbonate, and existing such structures become vulnerable to dissolution. Thus, ongoing acidification of the oceans also poses a threat to the food chains connected with the oceans. As members of the InterAcademy Panel, 105 science academies have issued a statement on ocean acidification. The statement recommends that by 2050, global CO_2 emissions be reduced by at least 50%, compared to the 1990 level.

Carbon Cycle

The carbon cycle describes the fluxes of carbon dioxide (CO_2) between the oceans, terrestrial biosphere, lithosphere, and the atmosphere. Human activities such as the combustion of fossil fuels and land use changes have led to a new flux of CO_2 into the atmosphere. About 45% has remained in the atmosphere; most of the rest has been taken up by the oceans, with some also taken up by terrestrial plants.

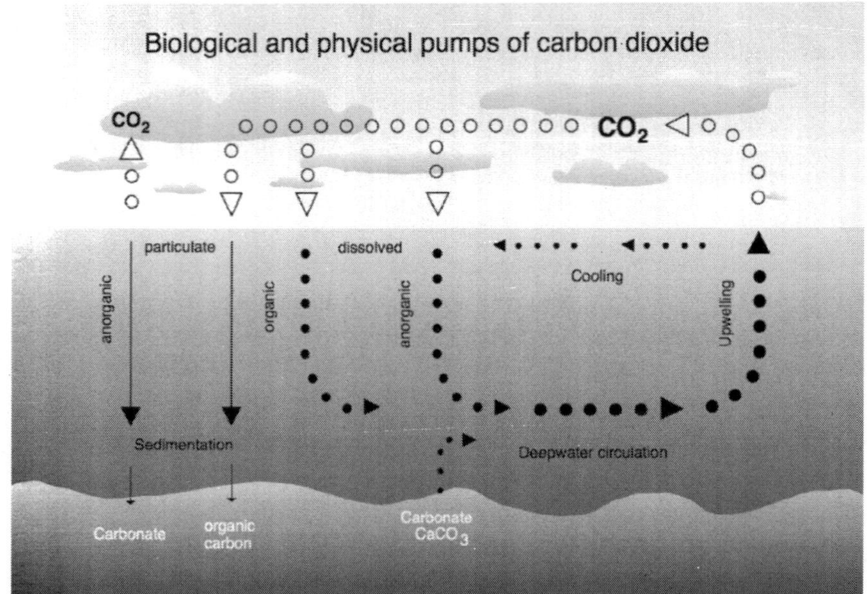

Figure: *The CO_2 cycle between the atmosphere and the ocean.*

The carbon cycle involves both organic compounds as well as inorganic carbon compounds such as carbon dioxide and the carbonates. The inorganic compounds are particularly relevant when discussing ocean acidification for it includes the many forms of dissolved CO_2 present in the Earth's oceans.

When CO_2 dissolves, it reacts with water to form a balance of ionic and non-ionic chemical species: dissolved free carbon dioxide (CO_2(aq)), carbonic acid (H_2CO_3), bicarbonate ($HCO-_3$) and carbonate (CO_2-_3). The ratio of these species depends on factors such as seawater temperature and alkalinity (as shown in a Bjerrum plot). These different forms of dissolved inorganic carbon are transferred from an ocean's surface to its interior by the ocean's solubility pump. The resistance of an area of ocean to absorbing atmospheric CO_2 is known as the Revelle factor.

Acidification

Dissolving CO_2 in seawater increases the hydrogen ion (H+) concentration in the ocean, and thus decreases ocean pH, as follows:

$$CO_{2\ (aq)} + H_2O \leftrightarrow H_2CO_3 \leftrightarrow HCO_3^- + H^+ \leftrightarrow CO_3^{2-} + 2\ H^+.$$

Caldeira and Wickett (2003) placed the rate and magnitude of modern ocean acidification changes in the context of probable historical changes during the last 300 million years.

Average surface ocean pH				
Time	***pH***	***pH change relative to pre-industrial***	***Source***	***H^+ concentration change relative to pre-industrial***
Pre-industrial (18th century)	8.179		analysed field	
Recent past (1990s)	8.104	–0.075	field	+ 18.9%
Present levels	~8.069	–0.11	field	+ 28.8%
2050 (2×CO 2 = 560 ppm)	7.949	–0.230	model	+ 69.8%
2100 (IS92a)	7.824	–0.355	model	+ 126.5%

Since the industrial revolution began, it is estimated that surface ocean pH has dropped by slightly more than 0.1 units on the logarithmic scale of pH, representing an approximately 29% increase in H+, and it is estimated that it will drop by a further 0.3 to 0.5 pH units (an additional doubling to tripling of today's post-industrial acid concentrations) by 2100 as the oceans absorb more anthropogenic CO_2, the impacts being most severe for coral reefs and the Southern Ocean. These changes are predicted to continue rapidly as the oceans take up more anthropogenic CO_2 from the atmosphere. The degree of change to ocean chemistry, including ocean pH, will depend on the mitigation and emissions pathways society takes.

Although the largest changes are expected in the future, a report from NOAA scientists found large quantities of water undersaturated in aragonite are already upwelling close to the Pacific continental shelf area of North America. Continental shelves play an important role in marine ecosystems since most marine organisms live or are spawned there, and though the study only dealt with the area from Vancouver to Northern California, the authors suggest that other shelf areas may be experiencing similar effects.

Rate

One of the first detailed datasets examining temporal variations in pH at a temperate coastal location found that acidification was occurring at a rate much higher than previously predicted, with consequences for near-shore benthic ecosystems. Thomas Lovejoy, former chief biodiversity advisor to the World Bank, has suggested that "the acidity of the oceans will more than double in the next 40 years. This rate is 100 times faster than any changes in ocean acidity in the last 20 million years, making it unlikely that marine life can somehow adapt to the changes." It is predicted that, by the year 2100, the level of acidity in the ocean will reach the levels experienced by the earth 20 million years ago.

Current rates of ocean acidification have been compared with the greenhouse event at the Paleocene–Eocene boundary (about 55 million years ago) when surface ocean temperatures rose by 5–6 degrees Celsius. No catastrophe was seen in surface ecosystems, yet bottom-dwelling organisms in the deep ocean experienced a major extinction. The current acidification is on a path to reach levels higher than any seen in the last 65 million years, and the rate of increase is about ten times the rate that preceded the Paleocene–Eocene mass extinction. The current and projected acidification has been described as an almost unprecedented geological event. A National Research Council study released in April 2010 likewise concluded that "the level of acid in the oceans is increasing at an unprecedented rate." A 2012 paper in the journal *Science* examined the geological record in an attempt to find a historical analog for current global conditions as well as those of the future. The researchers determined that the current rate of ocean acidification is faster than at any time in the past 300 million years.

A review by climate scientists at the RealClimate blog, of a 2005 report by the Royal Society of the UK similarly highlighted the centrality of the *rates* of change in the present anthropogenic acidification process, writing:

"The natural pH of the ocean is determined by a need to balance the deposition and burial of $CaCO_3$ on the sea floor against the influx of Ca_2+and CO_2–$_3$ into the ocean from dissolving rocks on land, called weathering. These processes stabilize the pH of the ocean, by a mechanism called $CaCO_3$ compensation... The point of bringing it up again is to note that if the CO_2 concentration of the atmosphere changes more slowly than this, as it always has throughout the Vostok record, the pH of the ocean will be relatively unaffected because $CaCO_3$ compensation can keep up. The [present] fossil fuel acidification is much faster than natural changes, and so the acid spike will be more intense than the earth has seen in at least 800,000 years."

In the 15-year period 1995–2010 alone, acidity has increased 6 percent in the upper 100 metres of the Pacific Ocean from Hawaii to Alaska. According to a statement in July 2012 by Jane Lubchenco, head of the U.S. National Oceanic and Atmospheric Administration "surface waters are changing much more rapidly than initial calculations have suggested. It's yet another reason to be very seriously concerned about the amount of carbon dioxide that is in the atmosphere now and the additional amount we continue to put out."

A 2013 study claimed acidity was increasing at a rate 10 times faster than in any of the evolutionary crises in the earth's history.

Calcification

Changes in ocean chemistry can have extensive direct and indirect effects on organisms and their habitats. One of the most important repercussions of increasing ocean acidity relates to the production of shells and plates out of calcium carbonate ($CaCO_3$). This process is called calcification and is important to the biology and survival of a wide range of marine organisms. Calcification involves the precipitation of dissolved ions into solid $CaCO_3$ structures, such as coccoliths. After they are formed, such structures are vulnerable to dissolution unless the surrounding seawater contains saturating concentrations of carbonate ions. The saturation state of seawater for a mineral (known as Ω) is a measure of the thermodynamic potential for the mineral to form or to dissolve, and is described by the following equation:

$$\Omega = \frac{\left[Ca^{2+}\right]\left[CO_3^{2-}\right]}{K_{sp}}$$

Here Ω is the product of the concentrations (or activities) of the reacting ions that form the mineral (Ca_2+and CO_2–$_3$), divided by the product of the concentrations of those ions when the mineral is at equilibrium (Ksp), that is, when the mineral is neither forming nor dissolving. In seawater, a natural horizontal boundary is formed as a result of temperature, pressure, and depth, and is known as the saturation horizon, or lysocline.

Above this saturation horizon, Ù has a value greater than 1, and $CaCO_3$ does not readily dissolve. Most calcifying organisms live in such waters. Below this depth, Ω has a value less than 1, and $CaCO_3$ will dissolve. However, if its production rate is high enough to offset dissolution, $CaCO_3$ can still occur where Ω is less than 1. The carbonate compensation depth occurs at the depth in the ocean where production is exceeded by dissolution.

As shown in the Bjerrum plot, along with the change in pH, adding extra CO_2 to the oceans also changes the oceans' concentrations of the different forms of dissolved inorganic carbon. There is a decrease in the concentration of CO_3^{2-}, which decreases Ω, and hence makes $CaCO_3$ dissolution more likely.

Calcium carbonate occurs in two common polymorphs: aragonite and calcite. Aragonite is much more soluble than calcite, with the result that the aragonite saturation horizon is always nearer to the surface than the calcite saturation horizon. This also means that those organisms that produce aragonite may possibly be more vulnerable to changes in ocean acidity than those that produce calcite. Increasing CO_2 levels and the resulting lower pH of seawater decreases the saturation state of $CaCO_3$ and raises the saturation horizons of both forms closer to the surface. This decrease in saturation state is believed to be one of the main factors leading to decreased calcification in marine organisms, as it has been found that the inorganic precipitation of $CaCO_3$ is directly proportional to its saturation state.

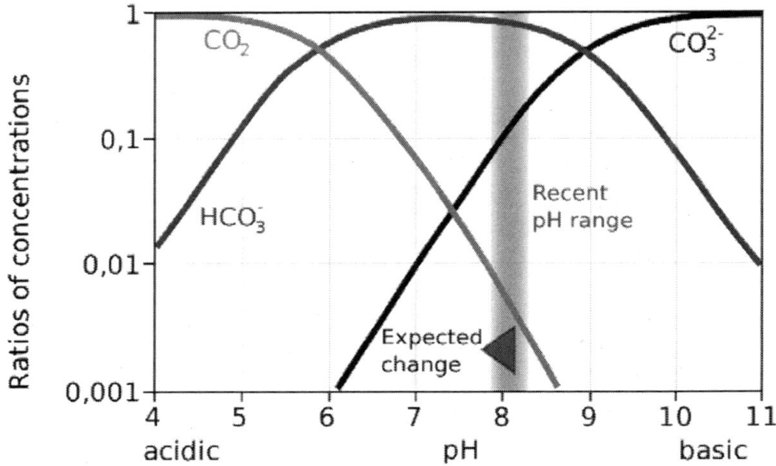

Figure: *Bjerrum plot: Change in carbonate system of seawater from ocean acidification.*

Possible Impacts

Impacts on Oceanic Calcifying Organisms: Although the natural absorption of CO_2 by the world's oceans helps mitigate the climatic effects of anthropogenic emissions of CO_2, it is believed that the resulting decrease in pH will have negative consequences, primarily for oceanic calcifying organisms. These span the food chain from autotrophs to heterotrophs and include organisms such as coccolithophores, corals, foraminifera, echinoderms, crustaceans and molluscs. As described above, under normal conditions, calcite and aragonite are stable in surface waters since the carbonate ion is at supersaturating concentrations. However, as ocean pH falls, the concentration of carbonate ions required for saturation to occur increases, and when carbonate becomes undersaturated, structures made of calcium carbonate are vulnerable to dissolution. Therefore, even if there is no change in the rate of calcification, the rate of dissolution of calcareous material increases.

Research has already found that corals, coccolithophore algae, coralline algae, foraminifera, shellfish and pteropods experience reduced calcification or enhanced dissolution when exposed to elevated CO_2.

The Royal Society published a comprehensive overview of ocean acidification, and its potential consequences, in June 2005. However, some studies have found different response to ocean acidification, with coccolithophore calcification and photosynthesis both increasing under elevated atmospheric pCO_2, an equal decline in primary production and calcification in response to elevated CO_2 or the direction of the response varying between species. A study in 2008 examining a sediment core from the North Atlantic found that while the species composition of coccolithophorids has remained unchanged for the industrial period 1780 to 2004, the calcification of coccoliths has increased by up to 40% during the same time. And another study in 2010 from Stony Brook University drew a dismal conclusion that while some areas are overharvested and other fishing grounds are being restored, because of ocean acidification it may be impossible to bring back many previous shellfish populations. While the full ecological consequences of these changes in calcification are still uncertain, it appears likely that many calcifying species will be adversely affected.

When exposed in experiments to pH reduced by 0.2 to 0.4, larvae of a temperate brittlestar, a relative of the common sea star, fewer than 0.1 percent survived more than eight days. There is also a suggestion that a decline in the coccolithophores may have secondary effects on climate, contributing to global warming by decreasing the Earth's albedo via their effects on oceanic cloud cover.

The fluid in the internal compartments where corals grow their exoskeleton is also extremely important for calcification growth. When the saturation rate of aragonite in the external seawater is at ambient levels, the corals will grow their aragonite crystals rapidly in their internal compartments, hence their exoskeleton grows rapidly. If the level of aragonite in the external seawater is lower than the ambient level, the corals have to work harder to maintain the right balance in the internal compartment. When that happens, the process of growing the crystals slows down, and this slows down the rate of how much their exoskeleton is growing. Depending on how much aragonite is in the surrounding water, the corals may even stop growing because the levels of aragonite are too low to pump in to the internal compartment. They could even dissolve faster than they can make the crystals to their skeleton, depending on the aragonite levels in the surrounding water.

Ocean acidification may also force some organisms to reallocate resources away from productive endpoints such as growth in order to maintain calcification.

Other Biological Impacts

Aside from the slowing and/or reversing of calcification, organisms may suffer other adverse effects, either indirectly through negative impacts on food resources, or directly as reproductive or physiological effects. For example, the elevated oceanic levels of CO_2 may produce CO_2-induced acidification of body fluids, known as hypercapnia. Also, increasing ocean acidity is believed to have a range of direct consequences.

For example, increasing acidity has been observed to: reduce metabolic rates in jumbo squid; depress the immune responses of blue mussels; and make it harder for juvenile clownfish to tell apart the smells of non-predators and predators, or hear the sounds of their predators. This is possibly because ocean acidification may alter the acoustic properties of seawater, allowing sound to propagate further, and increasing ocean noise.

This impacts all animals that use sound for echolocation or communication. A study performed by PLOS ONE concluded that Atlantic longfin squid eggs took longer to hatch in acidified water. Additionally, the squid statolith, an internal structure that helps them sense movement, was smaller and malformed in the squid placed in sea water with a lower pH.

However, as with calcification, as yet there is not a full understanding of these processes in marine organisms or ecosystems.

Shelled plankton species may flourish in altered oceans.

Nonbiological Impacts

Leaving aside direct biological effects, it is expected that ocean acidification in the future will lead to a significant decrease in the burial of carbonate sediments for several centuries, and even the dissolution of existing carbonate sediments. This will cause an elevation of ocean alkalinity, leading to the enhancement of the ocean as a reservoir for CO_2 with implications for climate change as more CO_2 leaves the atmosphere for the ocean.

Possible Responses

Reducing CO_2 Emissions: Members of the InterAcademy Panel recommended that by 2050, global anthropogenic CO_2 emissions be reduced less than 50% of the 1990 level. The 2009 statement also called on world leaders to:

- Acknowledge that ocean acidification is a direct and real consequence of increasing atmospheric CO_2 concentrations, is already having an effect at current concentrations, and is likely to cause grave harm to important marine ecosystems as CO_2 concentrations reach 450 [parts-per-million (ppm)] and above;
- [...] Recognise that reducing the build up of CO_2 in the atmosphere is the only practicable solution to mitigating ocean acidification;
- [...] Reinvigorate action to reduce stressors, such as overfishing and pollution, on marine ecosystems to increase resilience to ocean acidification.

Stabilizing atmospheric CO_2 concentrations at 450 ppm would require near-term emissions reductions, with steeper reductions over time.

The German Advisory Council on Global Change Stated: In order to prevent disruption of the calcification of marine organisms and the resultant risk of fundamentally altering marine food webs, the following guard rail should be obeyed: the pH of near surface waters should not drop more than 0.2 units below the pre-industrial average value in any larger ocean region (nor in the global mean).

One policy target related to ocean acidity is the magnitude of future global warming. Parties to the United Nations Framework Convention on Climate Change (UNFCCC) adopted a target of limiting warming to below 2 °C, relative to the pre-industrial level. Meeting this target would require substantial reductions in anthropogenic CO_2 emissions.

Limiting global warming to below 2 °C would imply a reduction in surface ocean pH of 0.16 from pre-industrial levels. This would represent a substantial decline in surface ocean pH.

Geoengineering

Geoengineering (mitigating temperature or pH effects of emissions) has been proposed as a possible response to ocean acidification. The IAP (2009) statement cautioned against geoengineering as a policy response:

Mitigation approaches such as adding chemicals to counter the effects of acidification are likely to be expensive, only partly effective and only at a very local scale, and may pose additional unanticipated risks to the marine environment.

There has been very little research on the feasibility and impacts of these approaches. Substantial research is needed before these techniques could be applied.

Reports by the WGBU (2006), the UK's Royal Society (2009), and the US National Research Council (2011) warned of the potential risks and difficulties associated with geoengineering.

Iron Fertilization

Iron fertilization of the ocean could stimulate photosynthesis in phytoplankton. The phytoplankton would convert the ocean's dissolved carbon dioxide into carbohydrate and oxygen gas, some of which would sink into the deeper ocean before oxidizing.

More than a dozen open-sea experiments confirmed that adding iron to the ocean increases photosynthesis in phytoplankton by up to 30 times. While this approach has been proposed as a potential solution to the ocean acidification problem, mitigation of surface ocean acidification might increase acidification in the less-inhabited deep ocean.

A report by the UK's Royal Society (2009) reviewed the approach for effectiveness, affordability, timeliness and safety. The rating for affordability was "medium", or "not expected to be very cost-effective." For the other three criteria, the ratings ranged from "low" to "very low" (i.e., not good).

For example, in regards to safety, the report found a "[high] potential for undesirable ecological side effects," and that ocean fertilization "may increase anoxic regions of ocean ('dead zones')."

Carbon Negative Fuels

Carbonic acid can be extracted from seawater as carbon dioxide for use in making synthetic fuel.

If the resulting flue exhaust gas was subject to carbon capture, then the process would be carbon negative over time, resulting in permanent extraction of inorganic carbon from seawater and the atmosphere with which seawater is in equilibrium. Based on the energy requirements, this process was estimated to cost about $50 per tonne of CO_2.

Oil Spill

An oil spill is the release of a liquid petroleum hydrocarbon into the environment, especially marine areas, due to human activity, and is a form of pollution.

The term is usually applied to marine oil spills, where oil is released into the ocean or coastal waters, but spills may also occur on land. Oil spills may be due to releases of crude oil from tankers, offshore platforms, drilling rigs and wells, as well as spills of refined petroleum products (such as gasoline, diesel) and their by-products, heavier fuels used by large ships such as bunker fuel, or the spill of any oily refuse or waste oil.

Spilt oil penetrates into the structure of the plumage of birds and the fur of mammals, reducing its insulating ability, and making them more vulnerable to temperature fluctuations and much less buoyant in the water.

Cleanup and recovery from an oil spill is difficult and depends upon many factors, including the type of oil spilled, the temperature of the water (affecting evaporation and biodegradation), and the types of shorelines and beaches involved. Spills may take weeks, months or even years to clean up.

Largest Oil Spills

Crude oil and refined fuel spills from tanker ship accidents have damaged natural ecosystems in Alaska, the Gulf of Mexico, the Galapagos Islands, France and many other places.

The quantity of oil spilled during accidents has ranged from a few hundred tons to several hundred thousand tons (e.g., Deepwater Horizon Oil Spill, Atlantic Empress, Amoco Cadiz) but is a limited barometre of damage or impact. Smaller spills have already proven to have a great impact on ecosystems, such as the Exxon Valdez oil spill because of the remoteness of the site or the difficulty of an emergency environmental response.

Oil spills at sea are generally much more damaging than those on land, since they can spread for hundreds of nautical miles in a thin oil slick which can cover beaches with a thin coating of oil. This can kill sea birds, mammals, shellfish and other organisms it coats. Oil spills on land are more readily containable if a makeshift earth dam can be rapidly bulldozed around the spill site before most of the oil escapes, and land animals can avoid the oil more easily.

Largest oil spills, ordered by tons					
Spill / Tanker	***Location***	***Date***	****Tons of crude oil (thousands)***	***Barrels (thousands)***	***US Gallons (thousands)***
Kuwaiti oil fires	Kuwait	January, 1991 - November, 1991	136,000-205,000	1,000,000-1,500,000	42,000,000-63,000,000
Kuwaiti oil lakes	Kuwait	January, 1991 - November, 1991	3,409-6,818	25,000-50,000	1,050,000-2,100,000
Lakeview Gusher	United States, Kern County, California	March 14, 1910 – September, 1911	1,200	9,000	378,000
Gulf War oil spill [d]	Kuwait, Iraq, and the Persian Gulf	January 19, 1991 - January 28, 1991	818–1,091	6,000–8,000	252,000–336,000
Deepwater Horizon	United States, Gulf of Mexico	April 20, 2010 – July 15, 2010	560-585	4,100-4,900	172,000-180,800
Ixtoc I	Mexico, Gulf of Mexico	June 3, 1979 – March 23, 1980	454–480	3,329–3,520	139,818–147,840
Atlantic Empress / Aegean Captain	Trinidad and Tobago	July 19, 1979	287	2,105	88,396
Fergana Valley	Uzbekistan	March 2, 1992	285	2,090	87,780
Nowruz Field Platform	Iran, Persian Gulf	February 4, 1983	260	1,907	80,080
ABT Summer	Angola, 700 nmi (1,300 km; 810 mi) offshore	May 28, 1991	260	1,907	80,080
Castillo de Bellver	South Africa, Saldanha Bay	August 6, 1983	252	1,848	77,616
Amoco Cadiz	France, Brittany	March 16, 1978	223	1,635	68,684

1. Jump up ^ One ton of crude oil is roughly equal to 308 US gallons or 7.33 barrels approx.; 1 oil barrel is equal to 35 imperial or 42 US gallons.
2. Jump up ^ Estimates for the amount of oil burned in the Kuwaiti oil fires range from 500,000,000 barrels (79,000,000 m^3) to nearly 2,000,000,000 barrels (320,000,000 m^3). 732 wells were set ablaze, while many others were severely damaged and gushed uncontrolled for several months. The fires alone were estimated to consume approximately 6,000,000 barrels (950,000 m^3) of oil per day at their peak. However, it is difficult to find reliable sources for the total amount of oil burned. The range of 1,000,000,000 barrels (160,000,000 m^3) to 1,500,000,000

barrels (240,000,000 m^3) given here represents frequently cited figures, but better sources are needed.

3. Jump up ^ Oil spilled from sabotaged fields in Kuwait during the 1991 Persian Gulf War pooled in approximately 300 oil lakes, estimated by the Kuwaiti Oil Minister to contain approximately 25,000,000 to 50,000,000 barrels (7,900,000 m^3) of oil. According to the U.S. Geological Survey, this does not include the amount of oil absorbed by the ground, forming a layer of "tarcrete" over approximately five percent of the surface of Kuwait, fifty times the area occupied by the oil lakes.
4. Jump up ^ Estimates for the Gulf War oil spill range from 4,000,000 to 11,000,000 barrels (1,700,000 m^3). The figure of 6,000,000 to 8,000,000 barrels (1,300,000 m^3) is the range adopted by the U.S. Environmental Protection Agency and the United Nations in the immediate aftermath of the war, 1991–1993, and is still current, as cited by NOAA and The New York Times in 2010. This amount only includes oil discharged directly into the Persian Gulf by the retreating Iraqi forces from January 19 to 28, 1991. However, according to the U.N. report, oil from other sources not included in the official estimates continued to pour into the Persian Gulf through June, 1991. The amount of this oil was estimated to be at least several hundred thousand barrels, and may have factored into the estimates above 8,000,000 barrels (1,300,000 m^3).

Human Impact

An oil spill represents an immediate fire hazard. The Kuwaiti oil fires produced air pollution that caused respiratory distress. The Deepwater Horizon explosion killed eleven oil rig workers. The fire resulting from the Lac-Mégantic derailment killed 47 and destroyed half of the town's centre.

Spilled oil can also contaminate drinking water supplies. For example, in 2013 two different oil spills contaminated water supplies for 300,000 in Miri, Malaysia; 80,000 people in Coca, Ecuador,. In 2000, springs were contaminated by an oil spill in Clark County, Kentucky.

Contamination can have an economic impact on tourism and marine resource extraction industries. For example, the Deepwater Horizon oil spill impacted beach tourism and fishing along the Gulf Coast, and the responsible parties were required to compensate economic victims.

Environmental Effects

Oil penetrates into the structure of the plumage of birds and the fur of mammals, reducing its insulating ability, and making them more vulnerable to temperature fluctuations and much less buoyant in the water.

Animals that rely on scent to find their babies or mothers fade away due to the strong scent of the oil. This causes a baby to be rejected and abandoned, leaving the babies to starve and eventually die. Oil can impair a bird's ability to fly, preventing it from foraging or escaping from predators. As they preen, birds may ingest the oil coating their feathers, irritating the digestive tract, altering liver function, and causing kidney damage.

Together with their diminished foraging capacity, this can rapidly result in dehydration and metabolic imbalance. Some birds exposed to petroleum also experience changes in their hormonal balance, including changes in their luteinizing protein. The majority of birds affected by oil spills die without human intervention. Some studies have suggested that less than one percent of oil-soaked birds survive, even after cleaning, although the survival rate can also exceed ninety percent, as in the case of the Treasure oil spill.

Heavily furred marine mammals exposed to oil spills are affected in similar ways. Oil coats the fur of sea otters and seals, reducing its insulating effect, and leading to fluctuations in body temperature and hypothermia. Oil can also blind an animal, leaving it defenceless. The ingestion of oil causes dehydration and impairs the digestive process. Animals can be poisoned, and may die from oil entering the lungs or liver.

There are three kinds of oil-consuming bacteria. Sulphate-reducing bacteria (SRB) and acid-producing bacteria are anaerobic, while general aerobic bacteria (GAB) are aerobic. These bacteria occur naturally and will act to remove oil from an ecosystem, and their biomass will tend to replace other populations in the food chain.

Cleanup and Recovery

Cleanup and recovery from an oil spill is difficult and depends upon many factors, including the type of oil spilled, the temperature of the water (affecting evaporation and biodegradation), and the types of shorelines and beaches involved.

Methods for cleaning up include:

- Bioremediation: use of microorganisms or biological agents to break down or remove oil; such as the bacteria Alcanivorax.

- Bioremediation Accelerator: Oleophilic, hydrophobic chemical, containing no bacteria, which chemically and physically bonds to both soluble and insoluble hydrocarbons. The bioremediation accelerator acts as a herding agent in water and on the surface, floating molecules to the surface of the water, including solubles such as phenols and BTEX, forming gel-like agglomerations. Undetectable levels of hydrocarbons can be obtained in produced water and manageable water columns. By overspraying sheen with bioremediation accelerator, sheen is eliminated within minutes. Whether applied on land or on water, the nutrient-rich emulsion creates a bloom of local, indigenous, pre-existing, hydrocarbon-consuming bacteria. Those specific bacteria break down the hydrocarbons into water and carbon dioxide, with EPA tests showing 98% of alkanes biodegraded in 28 days; and aromatics being biodegraded 200 times faster than in nature they also sometimes use the hydrofireboom to clean the oil up by taking it away from most of the oil and burning it.
- Controlled burning can effectively reduce the amount of oil in water, if done properly. But it can only be done in low wind, and can cause air pollution.

Figure: *Oil slicks on Lake Maracaibo.*

- Dispersants can be used to dissipate oil slicks. A dispersant is either a non-surface active polymer or a surface-active substance added to a suspension, usually a colloid, to improve the separation of particles and to prevent settling or clumping. They may rapidly disperse large amounts of certain oil types from the sea surface by transferring it into the water column. They will cause the oil slick to break up and form water-soluble micelles that are rapidly diluted. The oil is then effectively spread throughout a larger volume of water than the surface from where the oil was dispersed. They can also delay the formation of persistent oil-in-water emulsions. However, laboratory experiments showed that dispersants increased toxic hydrocarbon levels in fish by a factor of up to 100 and may kill fish eggs. Dispersed oil droplets infiltrate into deeper water and can lethally contaminate coral. Research indicates that some dispersants are toxic to corals. A 2012 study found that Corexit dispersant had increased the toxicity of oil by up to 52 times.
- Watch and wait: in some cases, natural attenuation of oil may be most appropriate, due to the invasive nature of facilitated methods of remediation, particularly in ecologically sensitive areas such as wetlands.
- Dredging: for oils dispersed with detergents and other oils denser than water.
- Skimming: Requires calm waters at all times during the process.
- Solidifying: Solidifiers are composed of dry hydrophobic polymers that both adsorb and absorb. They clean up oil spills by changing the physical state of spilled oil from liquid to a semi-solid or a rubber-like material that floats on water. Solidifiers are insoluble in water, therefore the removal of the solidified oil is easy and the oil will not leach out. Solidifiers have been proven to be relatively non-toxic to aquatic and wild life and have been proven to suppress harmful vapours commonly associated with hydrocarbons such as Benzene, Xylene, Methyl Ethyl, Acetone and Naphtha. The reaction time for solidification of oil is controlled by the surf area or size of the polymer as well as the viscosity of the oil. Some solidifier product manufactures claim the solidified oil can be disposed of in landfills, recycled as an additive in asphalt or rubber products, or burned as a low ash fuel. A solidifier called C.I.Agent (manufactured by C.I.Agent Solutions of Louisville, Kentucky) is being used by BP in granular form, as well as in Marine and Sheen Booms at Dauphin Island

and Fort Morgan, Alabama, to aid in the Deepwater Horizon oil spill cleanup.

- Vacuum and centrifuge: oil can be sucked up along with the water, and then a centrifuge can be used to separate the oil from the water - allowing a tanker to be filled with near pure oil. Usually, the water is returned to the sea, making the process more efficient, but allowing small amounts of oil to go back as well. This issue has hampered the use of centrifuges due to a United States regulation limiting the amount of oil in water returned to the sea.

Equipment used includes:

- Booms: large floating barriers that round up oil and lift the oil off the water
- Skimmers: skim the oil
- Sorbents: large absorbents that absorb oil
- Chemical and biological agents: helps to break down the oil
- Vacuums: remove oil from beaches and water surface
- Shovels and other road equipment: typically used to clean up oil on beaches

Prevention

- Secondary containment - methods to prevent releases of oil or hydrocarbons into environment.
- Oil Spill Prevention Containment and Countermeasures (SPCC) programme by the United States Environmental Protection Agency.
- Double-hulling - build double hulls into vessels, which reduces the risk and severity of a spill in case of a collision or grounding. Existing single-hull vessels can also be rebuilt to have a double hull.
- Thick-hulled railroad transport tanks

Environmental Sensitivity Index (ESI) Mapping

Environmental Sensitivity Index (ESI) maps are used to identify sensitive shoreline resources prior to an oil spill event in order to set priorities for protection and plan cleanup strategies. By planning spill response ahead of time, the impact on the environment can be minimized or prevented. Environmental sensitivity index maps are basically made up of information within the following three categories: shoreline type, and biological and human-use resources.

Shoreline Type

Shoreline type is classified by rank depending on how easy the garet would be to clean up, how long the oil would persist, and how sensitive the shoreline is. The floating oil slicks put the shoreline at particular risk when they eventually come ashore, covering the substrate with oil. The differing substrates between shoreline types vary in their response to oiling, and influence the type of cleanup that will be required to effectively decontaminate the shoreline. In 1995, the US National Oceanic and Atmospheric Administration extended ESI maps to lakes, rivers, and estuary shoreline types.

The exposure the shoreline has to wave energy and tides, substrate type, and slope of the shoreline are also taken into account—in addition to biological productivity and sensitivity. The productivity of the shoreline habitat is also taken into account when determining ESI ranking. Mangroves and marshes tend to have higher ESI rankings due to the potentially long-lasting and damaging effects of both the oil contamination and cleanup actions. Impermeable and exposed surfaces with high wave action are ranked lower due to the reflecting waves keeping oil from coming onshore, and the speed at which natural processes will remove the oil.

Biological Resources

Habitats of plants and animals that may be at risk from oil spills are referred to as "elements" and are divided by functional group. Further classification divides each element into species groups with similar life histories and behaviours relative to their vulnerability to oil spills. There are eight element groups: Birds, Reptiles, Amphibians, Fish, Invertebrates, Habitats and Plants, Wetlands, and Marine Mammals and Terrestrial Mammals. Element groups are further divided into sub-groups, for example, the 'marine mammals' element group is divided into dolphins, manatees, pinnipeds (seals, sea lions & walruses), polar bears, sea otters and whales. Problems taken into consideration when ranking biological resources include the observance of a large number of individuals in a small area, whether special life stages occur ashore (nesting or molting), and whether there are species present that are threatened, endangered or rare.

Human-use Resources

Human use resources are divided into four major classifications; archaeological importance or cultural resource site, high-use recreational areas or shoreline access points, important protected management areas, or resource origins. Some examples include airports, diving sites, popular beach sites, marinas, natural reserves or marine sanctuaries.

Estimating the Volume of a Spill

By observing the thickness of the film of oil and its appearance on the surface of the water, it is possible to estimate the quantity of oil spilled. If the surface area of the spill is also known, the total volume of the oil can be calculated.

	Film thickness			*Quantity spread*	
Appearance	*in*	*mm*	*nm*	*gal/sq mi*	*L/ha*
Barely visible	0.0000015	0.0000380	38	25	0.370
Silvery sheen	0.0000030	0.0000760	76	50	0.730
First trace of colour	0.0000060	0.0001500	150	100	1.500
Bright bands of colour	0.0000120	0.0003000	300	200	2.900
Colours begin to dull	0.00004	0.0010000	1000	666	9.700
Colours are much darker	0.0000800	0.0020000	2000	1332	19.500

Oil spill model systems are used by industry and government to assist in planning and emergency decision making. Of critical importance for the skill of the oil spill model prediction is the adequate description of the wind and current fields. There is a worldwide oil spill modelling (WOSM) programme. Tracking the scope of an oil spill may also involve verifying that hydrocarbons collected during an ongoing spill are derived from the active spill or some other source. This can involve sophisticated analytical chemistry focused on finger printing an oil source based on the complex mixture of substances present. Largely, these will be various hydrocarbons, among the most useful being polyaromatic hydrocarbons. In addition, both oxygen and nitrogen heterocyclic hydrocarbons, such as parent and alkyl homologues of carbazole, quinoline, and pyridine, are present in many crude oils. As a result, these compounds have great potential to supplement the existing suite of hydrocarbons targets to fine tune source tracking of petroleum spills. Such analysis can also be used to follow weathering and degradation of crude spills.

Marine Fuel Management

Marine fuel management (MFM) is a multi-level approach to measuring, monitoring, and reporting fuel usage on a boat or ship, with the goals of reducing fuel usage, increasing operational efficiency, and improving fleet management oversight. MFM has grown in importance due to the rising costs of marine fuel and increased governmental pressures to reduce the pollution generated by the world's fleet.

Effective MFM requires that you know:

- How much fuel is used
- How the fuel was used
- What things impact fuel usage
- And by how much

Manual methods of measuring fuel usage, i.e. fuel tank dipping or sounding, typically do not tell how much fuel was used:

- Travelling versus idling while in port or on station
- By a specific engine (port versus starboard, for example)
- Performing one job versus another
- By crew A versus crew B on similar voyages

Without a clear understanding of how fuel is being used, there is no operational baseline from which to compare any kind of fuel conservation tool or activity. Without a baseline, there is no way to determine if conservation strategies are actually working.

MFM allows a fleet owner to track actual fuel consumption and relate fuel consumption to the work performed by the vessel. It supports the analysis of the effectiveness of operating strategies and helps develop a clearer understanding of how well a vessel uses its fuel.

MFM Main Functional Areas

- Operational Performance
- Engineering and Maintenance Management
- Management Oversight

Operational Performance

Operational performance includes those functional areas that impact the actual performance of a vessel or fleet. It includes fuel monitoring, inventory control, accounting, and engine throttle management.

Fuel Monitoring

Many marine vessels do not provide a way for captain and crew to measure and monitor fuel usage while underway. An optimum system onboard would include the ability to instantaneously monitor fuel burn rates from the wheelhouse. Individual engine and generator burn rates would be included, as well as fuel tank levels. This proactive monitoring would allow the crew to make decisions that positively impact fuel burn rates and efficiency.

Inventory Control

Fuel tanks need to have sensors installed that continuously monitor levels as fuel is taken onboard and burned by engines and generators. Periodically measuring tank levels using traditional manual methods is not accurate enough or timely, given the volumes of fuel that a marine engine can consume. Metres should be installed on transfer lines where fuel is taken onboard or off-loaded.

Accounting

In some parts of the world, fuel theft is an ongoing concern. Consequently, the accurate measurement of fuel taken on board coupled with the fuel actually consumed by engines and generators, is an important part of MFM. Metres should be installed in all fuel transfer lines so accurate fuelling data can be captured. This data can then be compared with burn rates to determine whether fuel is being transferred off the vessel secretively. Beyond fuel theft, many governmental jurisdictions require that all fuel spill incidents be recorded and reported to the local authorities. For example, the Marine Department of the Government of Hong Kong has specific guidelines for responding to accidental marine fuel spills which reflect international requirements as promulgated by MARPOL, the International Convention for the Prevention of Pollution From Ships.

Additionally, accounting for fuel usage at various points along a voyage provides the ability to tie fuel burn and its associated costs to shipping or container rates. For example, understanding how a vessel burns fuel on certain parts of a voyage, allows the more accurate bidding of container rates so profit margins stay healthy. Consequently, varying shipping rates based on documented fuel usage rates can allow a shipper to bid more aggressively. A modern marine fuel management system would help in monitoring fuel usage, fuel transfers, bunkering events and could be configured to sound an audible alarm when refilling fuel tanks might lead to a spill.

Throttle Management

Vessel operators have the most control over fuel usage by the way they use the engine(s) throttle. Wind, current, hull condition, load, and propulsion system health can all impact fuel burn both positively or negatively. Some operators choose to lower engine speed, and hence vessel speed, in an attempt to save fuel. However, engine RPM and vessel speed alone are not indicative of total fuel consumption, so arbitrarily lowering engine speed does not guarantee fuel savings. One must do the workflow calculations on how the propulsion system is operating under existing changing conditions and then tie that to fuel consumption. Simply lowering engine RPM does not guarantee an optimum vessel speed setting based on conditions. Some modern fuel management systems are designed to perform these calculations while underway and make recommendations to the vessel master.

Engineering and Maintenance Management

As with any capital asset, manufacturers typically include the standard maintenance practices and procedures needed to keep the asset functioning properly and within design specifications. In many cases, scheduled maintenance routines are based on laboratory or design parameters and do not necessarily represent the optimum. MFM supports proper maintenance on marine engines and generators by using the actual fuel burned or hours operated as the basis for performing maintenance routines. This condition-based maintenance programme more accurately reflects the operating environment of the engine, but more importantly, reduces or eliminates unnecessary maintenance work.

Management Oversight

Management functions within MFM include:

- Vessel performance analysis and overall fleet performance
- Crew analysis with emphasis on applying lessons learned as best practice across the fleet
- Key Performance Indicator (KPI) gathering across the fleet to include fuel burned per mile or per ton; throttle settings at various points on a voyage; engine RPMs and exhaust gas analysis; and vessel performance against hull conditions.
- Fuel management from purchasing to transfers to usage
- Chartered vessel fuel performance and adherence to contractual obligations

Coastal Plain

A coastal plain is an area of flat, low-lying land adjacent to a seacoast and separated from the interior by other features. One of the world's longest coastal plains is located in eastern South America. The Gulf Coastal Plain of North America extends northwards from the

Gulf of Mexico along the Lower Mississippi River to the Ohio River, which is a distance of about 500 miles (about 800 km). During the Cretaceous period, the central area of the United States was covered by a shallow sea, which disappeared as the land rose. Large fossilized aquatic birds called *Hesperornis* and *Ichthyornis*, found in western Kansas, indicate that the shallow sea was rife with fish.

Continental Margin

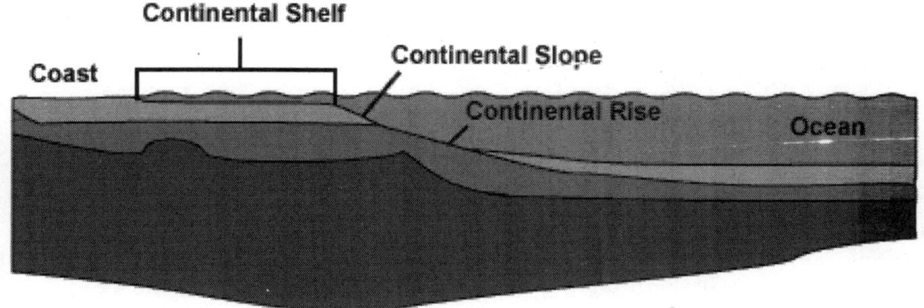

Figure: *Profile illustrating the shelf, slope and rise*

The continental margin is the zone of the ocean floor that separates the thin oceanic crust from thick continental crust. Together, the continental shelf, continental slope, and continental rise are called continental margin. Continental margins constitute about 28% of the oceanic area.

The transition from continental to oceanic crust commonly occurs within the outer part of the margin, called continental rise. Oceanwards beyond the edge of the rise lies the abyssal plain. The underwater part of the continental crust is called continental shelf, which usually abruptly terminates with the continental slope, which in turn terminates with the foot of the slope. The under-ocean part constitutes about 20% of the continental crust.

The edge of the continental margin is one criterion for the boundary of the internationally recognized claims to underwater resources by countries in the definition of the "Continental Shelf" by the United Nations Convention on the Law of the Sea (although in the UN definition the "legal continental shelf" may extend beyond the geomorphological continental shelf and vice versa).

Earthquakes and volcanic eruptions are common on the west coast of North America, but there are no active volcanoes on the east coast and large magnitude earthquakes are rare. Geologists refer to the west coast as a (tectonically) active continental margin and the east coast a passive continental margin. These differences can be found on several continents. Most passive continental margins have broad continental shelves, whereas active continental margins typically have narrow continental shelves. In addition, active continental margins are near plate boundaries whereas passive continental margins are further away from an active plate boundary.

Continental Shelf

The continental shelf is the extended perimeter of each continent and associated coastal plain. Much of the shelf was exposed during glacial periods, but it is now submerged under relatively shallow seas (known as shelf seas) and gulfs and was similarly submerged during other interglacial periods.

The continental margin, between the continental shelf and the abyssal plain, comprises a steep continental slope followed by the flatter continental rise. Sediment from the continent above cascades down the slope and accumulates as a pile of sediment at the base of the slope, called the continental rise. Extending as far as 500 km from the slope, it consists of thick sediments deposited by turbidity currents from the shelf and slope. The continental rise's gradient is intermediate between the slope and the shelf, on the order of 0.5–1°. Under the United Nations Convention on the Law of the Sea, the name continental shelf was given a legal definition as the stretch of the seabed adjacent to the shores of a particular country to which it belongs.

Geographical Distribution

The width of the continental shelf varies considerably – it is not uncommon for an area to have virtually no shelf at all, particularly where the forward edge of an advancing oceanic plate dives beneath continental crust in an offshore subduction zone such as off the coast of Chile or the west coast of Sumatra. The largest shelf – the Siberian Shelf in the Arctic Ocean – stretches to 1,500 kilometers (930 mi) in width. The South China Sea lies over another extensive area of continental shelf, the Sunda Shelf, which joins Borneo, Sumatra, and Java to the Asian mainland. Other familiar bodies of water that overlie continental shelves are the North Sea and the Persian Gulf. The average width of continental shelves is about 80 km (50 mi). The depth of the shelf also varies, but is generally limited to water shallower than 150 m (490 ft). The slope of the shelf is usually quite low, on the order of 0.5°; vertical relief is also minimal, at less than 20 m (66 ft).

Though the continental shelf is treated as a physiographic province of the ocean, it is not part of the deep ocean basin proper, but the flooded margins of the continent. Passive continental margins such as most of the Atlantic coasts have wide and shallow shelves, made of thick sedimentary wedges derived from long erosion of a neighbouring continent. Active continental margins have narrow, relatively steep shelves, due to frequent earthquakes that move sediment to the deep sea.

Topography

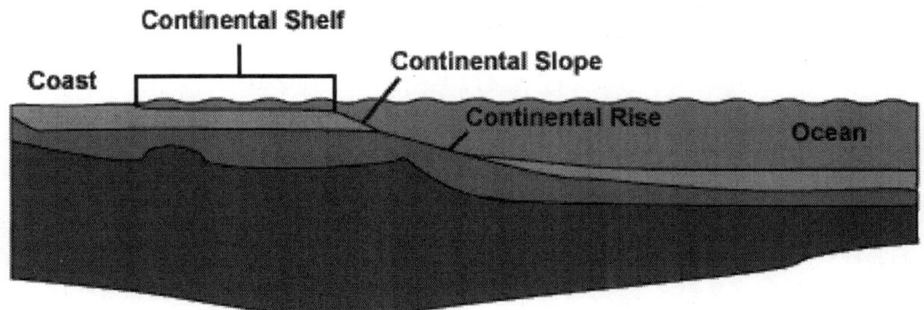

The shelf usually ends at a point of increasing slope (called the *shelf break*). The sea floor below the break is the *continental slope.* Below the slope is the *continental rise*, which finally merges into the deep ocean floor, the abyssal plain. The continental shelf and the slope are part of the continental margin.

The shelf area is commonly subdivided into the *inner continental shelf*, *mid continental shelf*, and *outer continental shelf*, each with their specific geomorphology and marine biology.

The character of the shelf changes dramatically at the shelf break, where the continental slope begins. With a few exceptions, the shelf break is located at a remarkably uniform depth of roughly 140 m (460 ft); this is likely a hallmark of past ice ages, when sea level was lower than it is now.

The continental slope is much steeper than the shelf; the average angle is 3°, but it can be as low as 1° or as high as 10°. The slope is often cut with submarine canyons. The physical mechanisms involved in forming these canyons were not well understood until the 1960s.

Sediments

The continental shelves are covered by terrigenous sediments; that is, those derived from erosion of the continents. However, little of the sediment is from current rivers; some 60-70% of the sediment on the world's shelves is *relict sediment*, deposited during the last ice age, when sea level was 100–120 m lower than it is now.

Sediments usually become increasingly fine with distance from the coast; sand is limited to shallow, wave-agitated waters, while silt and clays are deposited in quieter, deep water far offshore. These shelf sediments accumulate at an average rate of 30 cm/1000 years, with a range from 15–40 cm. Though slow by human standards, this rate is much faster than that for deep-sea pelagic sediments.

Biota

Continental shelves teem with life, because of the sunlight available in shallow waters, in contrast to the biotic desert of the oceans' abyssal plain. The pelagic (water column) environment of the continental shelf constitutes the neritic zone, and the benthic (sea floor) province of the shelf is the sublittoral zone.

Though the shelves are usually fertile, if anoxic conditions prevail during sedimentation, the deposits may over geologic time become sources for fossil fuels.

Economic Significance

The relatively accessible continental shelf is the best understood part of the ocean floor. Most commercial exploitation from the sea, such as metallic-ore, non-metallic ore, and hydrocarbon extraction, takes place on the continental shelf.

Sovereign rights over their continental shelves up to a depth of 200 metres or to a distance where the depth of waters admitted of resource exploitation were claimed by the marine nations that signed the Convention on the Continental Shelf drawn up by the UN's International Law Commission in 1958.

This was partly superseded by the 1982 United Nations Convention on the Law of the Sea. which created the 200 nautical mile exclusive economic zone and extended continental shelf rights for states with physical continental shelves that extend beyond that distance.

The legal definition of a continental shelf differs significantly from the geological definition. UNCLOS states that the shelf extends to the limit of the continental margin, but no less than 200 nautical miles from the baseline.

Thus inhabited volcanic islands such as the Canaries, which have no actual continental shelf, nonetheless have a legal continental shelf, whereas uninhabitable islands have no shelf.

Flat Coast

At a flat coast or flat shoreline, the land descends gradually into the sea. Flat coasts can be formed either as a result of the sea advancing into gently-sloping terrain or through the abrasion of loose rock. They may be basically divided into two parallel strips: the shoreface and the beach.

Flat coasts consist of loose material such as sand and gravel. Wind transports finer grains of sand inland over the dunes. The sea washes pebbles and sand away from the coast and dumps it at other locations.

Flat Coast Littoral Series

The typical sequence of landforms created by the sea is described as a "littoral series".

Sandbars, Runnels and Creeks

The littoral series of a flat coast starts in the permanently flooded shallow water region, or shoreface, with a sand or gravel reef (also called a bar). The longshore bar is an elongated ridge of sand found parallel to the shore in the surf zone on many flat coasts. It consists mainly of sand or gravel, depending on the material available along the coast. The sides of the sandbar fall gently away.

The basin between a sandbar and the shore zone is called the runnel or swale. The presence of a bar clearly indicates that the movement of waves is transporting and depositing material on the seabed. There may be several bars whose longitudinal axes all run parallel to the beach and which are separated by equally parallel runnels or creeks. The drainage troughs in areas of tidal flats also run parallel to the coast.

Shoreface and Beach

The shoreface (or underwater platform) on flat coasts encompasses in its narrow sense that area which is subject to the constant action of moving water. This means that the landward boundary between shoreface and beach is the line of the average low-water mark. However this definition is not universal and frequently varies from author to author in the literature. Whilst some define the beach as the landward transition to the shoreface that extends from the low-water mark to the highest high-water mark, i.e. the zone that is only periodically or episodically (after a storm surge) flooded by water; other authors do not use the term "beach" for the landward element of a flat coast at all.

Figure: *Start of the dune belt by the sand cliff*

They describe the region between the mean low-water mark and the mean high-water mark of the tides as the intertidal zone or foreshore and that area above the average high-water mark as the supratidal zone or backshore that is only directly attacked by water during storms.

Because the backshore is often considerably flatter in appearance than the foreshore, which slopes clearly down towards the sea, it is also often referred to as a beach platform, which is why this part of the shore can be considered in practice to be the actual beach. The farthest point inland that is reached by storm surges is bounded by a belt of dunes, where floods can form a dune cliff.

Berm

On the beach (the beach platform) there is very often a bank of sand or a gravel ridge parallel to the shoreline and a few tens of centimetres high, known as the berm. On its landward side there is often a shallow runnel. The berm is formed by material transported by the breaking waves that is thrown beyond the average level of the sea.

The coarse-grained material that can no longer be washed away by the backwash remains behind. The location and size of the berm is subject to seasonal changes. For example, a winter bern that has been thrown up by storm surges in winter is usually much more prominent and higher up the beach than berms formed by summer high tides.

Beach Losses and Gains

Beaches are usually heavily eroded during storm surges and the beach profile steepened, whereas normal wave action on flat coasts tends to raise the beach. Not infrequently a whole series of parallel berms is formed, one behind the other.

There is a consequent gradual increase in height with the result that, over time, the shoreline advances seawards. A striking example of land-forming system of berms is Skagen Odde on the northern tip of Vendsyssel in the extreme north of Denmark. This headland is still growing today as more berms are added.

Coastal defences against erosion are groynes, stone walls, or tetrapods of concrete, which act as breakwaters. The first plants to colonise the dunes include sea buckthorn or beach grass which prevent wind erosion.

Natural Arch

A natural arch or natural bridge is a natural rock formation where a rock arch forms, with an opening underneath. Most natural arches form as a narrow bridge, walled by cliffs, become narrower from erosion, with a softer rock stratum under the cliff-forming stratum gradually eroding out until the rock shelters thus formed meet underneath the ridge, thus forming the arch.

Natural arches commonly form where cliffs are subject to erosion from the sea, rivers or weathering (subaerial processes); the processes "find" weaknesses in rocks and work on them, making them larger until they break through.

The choice between *bridge* and *arch* is somewhat arbitrary. The Natural Arch and Bridge Society identifies a bridge as a subtype of arch that is primarily water-formed. By contrast, the *Dictionary of Geological Terms* defines a natural bridge as a "natural arch that spans a valley of erosion."

Coastline

On coasts two different types of arches can form depending on the geology. On discordant coastlines rock types run at 90° to the coast. Wave refraction concentrates the wave energy on the headland,

and an arch forms when caves break through the headland, e.g., London Bridge in (Victoria, Australia). When these eventually collapse, they form stacks and stumps.

On concordant coastlines rock types run parallel to the coastline, with weak rock (such as shale) protected by stronger rock (such as limestone) the wave action breaks through the strong rock and then erodes the weak rock very quickly. Good examples of this are at Durdle Door and Stair Hole near Lulworth Cove on the Dorset Jurassic Coast in south England, although these are on an area of concordant coastline. When Stair Hole eventually collapses, it will form a cove.

Weather-eroded Arches

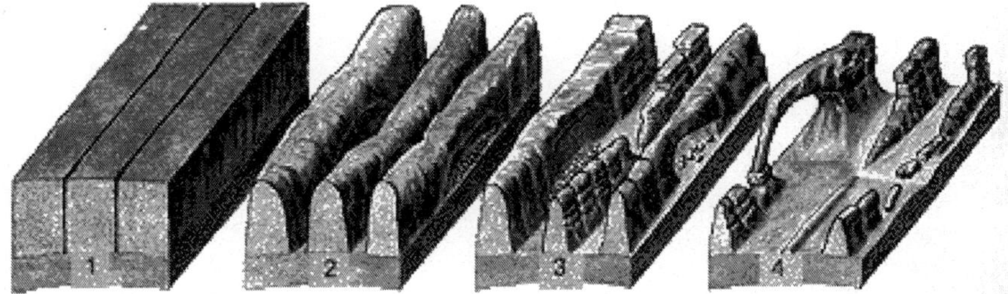

Figure: *A diagram showing the sequence of arch formation.*

1. Deep cracks penetrate into a sandstone layer.
2. Erosion wears away exposed rock layers and enlarges the surface cracks, isolating narrow sandstone walls, or fins.
3. Alternating frosts and thawing cause crumbling and flaking of the porous sandstone and eventually cut through some of the fins.
4. The resulting holes become enlarged to arch proportions by rockfalls and weathering. Arches eventually collapse, leaving only buttresses that in time will erode.

- Many of these arches are found within Arches National Park and Rainbow Bridge National Monument in Utah.

Water-eroded Arches

Some natural bridges may look like arches, but they form in the path of streams that wear away and penetrate the rock. Pothole arches form by chemical weathering as water collects in natural depressions and eventually cuts through to the layer below. Natural Bridges National Monument in Utah is another area to view several natural bridges.

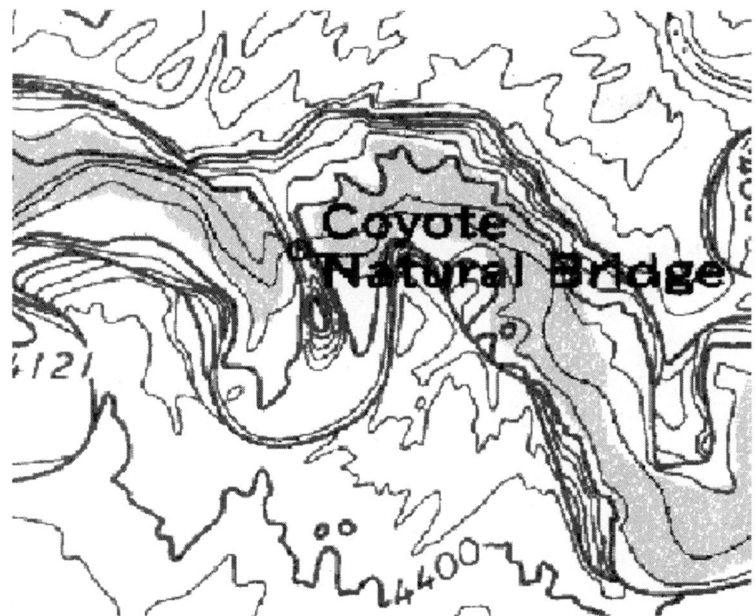

Figure: *A topographic map of Coyote Natural Bridge in Utah shows how the meandering Coyote Gulch carved a shorter route through the rock under the arch. The old riverbed is now higher than the present water level.*

Cave Erosion

Natural bridges can form from natural limestone caves, where paired sinkholes collapse and a ridge of stone is left standing in between, with the cave passageway connecting from sinkhole to sinkhole.

Like all rock formations, natural bridges are subject to continued erosion, and will eventually collapse and disappear. One example of this was the double-arched Victorian coastal rock formation, London Bridge, which lost an arch after storms increased erosion.

Arches as Highway or Railway Bridges

In a few places in the world, natural arches are truly natural bridges because there are roads or railroads running across them.

In Virginia, US Route 11 traverses the famous Natural Bridge.

Two additional such arches are found in Kentucky. One, a cave erosion arch made of limestone, is located in Carter Caves State Park and it has a paved road on top. Another, a weather-eroded sandstone arch with a dirt road on top is located on the edge of Natural Bridge State Resort Park in Kentucky. It is called White's Branch Arch (also known as the Narrows) and the road going over it is usually referred to as the Narrows Road. Another is found in Ponoarele Village, in Romania. It is 60 m long, 13 m wide, features a stone arch 4 m thick, 20 m high, with a 9 m span. It is called God's Bridge (Podul lui Dumnezeu) and it is effectively used for traffic.

The railroad from Lima, Peru crosses the Rio Yauli on a natural bridge near kilometer 214.2 as it approaches the city of La Oroya, Peru

Tidal Marsh

A tidal marsh is a type of marsh that is found along coasts and estuaries of which the flooding characteristics are determined by the tidal movement of the adjacent estuary, sea or ocean. According to the salinity of the flooding water, freshwater, brackish and saline tidal marshes are distinguished. Respectively, they may be classified into coastal marshes and estuarine marshes.

They are also commonly zoned into lower marshes (also called intertidal marshes) and upper or high marshes, based on their elevation with respect to the sea level.

In addition they may also be classified into back-barrier marshes, estuarine brackish marshes and tidal freshwater marshes, according to the degree of the influence of the sea level.

Island and Barrier Island

Tidal Marshes can contain island off their shore called barrier islands. These cigar shaped islands form parallel and close to the shoreline of a tidal marsh. The islands fully exposed during low tide look like a hill, and during a high tide being fully surrounded by water look like an island. Formation of barrier islands have been explained in many ways by history's scientific minds, some mechanisms of barrier island formation are offshore bar theory, spit accretion theory, and formation can even be due to climate change. Contrary to what one might think the presence of saline resistant plants does not affect the erosion rates of tidal marsh islands, rather the contributing factor to erosion rates is the soil type of the island.

Tide Pool

Tide pools, or rock pools, are rocky pools by oceans that are filled with seawater. Many of these pools exist as separate entities only at low tide.

Tide pools are habitats of uniquely adaptable animals that have engaged the special attention of naturalists and marine biologists, as well as philosophical essayists: John Steinbeck wrote in *The Log from the Sea of Cortez*, "It is advisable to look from the tide pool to the stars and then back to the tide pool again."

Zones from Shallow to Deep

Figure: *Tide pools in Santa Cruz, California from spray/splash zone to low tide zone*

Tidal pools exist in the intertidal zones. These zones receive spray from wave action during high tides and storms. At other times the rocks experience other extreme conditions, baking in the sun or exposed to cold winds. Few organisms can survive such harsh conditions. Lichens and barnacles live in this region. In this zone, different barnacle species live at very tightly constrained elevations. Tidal conditions precisely determine the exact height of an assemblage relative to sea level.

The intertidal zone is periodically exposed to sun and wind, so it desiccates barnacles which need to be well adapted to water loss. Their calcite shells are impermeable, and they possess two plates which they slide across their mouth opening when not feeding. These plates also protect against predation.

High Tide Zone

The high tide zone is flooded during each high tide. Organisms must survive wave action, currents, and exposure to the sun. This zone is predominantly inhabited by seaweed and invertebrates, such as sea anemones, seastars, chitons, crabs, green algae, and mussels. Marine algae provide shelter for nudibranchs and hermit crabs.

The same waves and currents that make the life in the high tide zone so difficult bring food to filter feeders and other intertidal organisms.

Figure: *Low tide zone in a tide pool*

Low Tide Zone

This subregion is mostly submerged, and is exposed only during low tide. It teems with life and has much more marine vegetation, especially seaweeds. There is also greater biodiversity. Organisms in this zone do not have to be as well adapted to drying out and temperature extremes. Low tide zone organisms include abalone, anemones, brown seaweed, chitons, crabs, green algae, hydroids, isopods, limpets, mussels, nudibranchs, sculpin, sea cucumber, sea lettuce, sea palms, sea stars, sea urchins, shrimp, snails, sponges, surf grass, tube worms, and whelks.

These creatures can grow to larger sizes because there is more available energy and better water coverage: The water is shallow enough to allow more light for photosynthetic activity, and the salinity is at almost normal levels. This area is also relatively protected from large predators because of the wave action and shallow water.

Life in the Tide Pool

Tide pools provide a home for hardy organisms such as sea stars, mussels and clams. Inhabitants must be able to cope with a constantly changing environment — fluctuations in water temperature, salinity, and oxygen content. Huge waves, strong currents, exposure to midday sun and predators are only a few of the hazards that tide pools' animals must endure to survive. Waves can dislodge mussels and draw them out to sea. Gulls pick up and drop sea urchins to break them open. Starfish prey on mussels and are eaten by gulls themselves. Even black bears sometimes feast on intertidal creatures at low tide. Although tide pool organisms must avoid getting washed away into the ocean, drying up in the sun, or getting eaten, they depend on the tide pool's constant changes for food.

Fauna

The sea anemone *Anthopleura elegantissima* produces clones of itself in order to reproduce through a process called longitudinal fission, in which the animal splits into two parts along its length. The sea anemone *Anthopleura sola* often engages in territorial fights. The white tentacles (acrorhagi), which contain stinging cells, are for fighting. The sea anemones sting each other repeatedly until one moves. Some species of starfish have the ability to regenerate lost arms. Most species must retain an intact central part of the body to be able to regenerate, but a few can regrow from a single ray. The regeneration of these stars is possible because the vital organs are in the arms.

Flora

Sea palms look very much as palm trees do. They live in the middle to upper intertidal zones in areas with greater wave action. High wave action may increase nutrient availability and moves the blades of the thallus, allowing more sunlight to reach the organism so that it can photosynthesize. In addition, the constant wave action removes competitors, such as the mussel species *Mytilus californianus*. Recent studies have shown that *Postelsia* grows in greater numbers when such competition exists — a control group with no competition produced fewer offspring than an experimental group with mussels; from this it is thought that the mussels provide protection for the developing gametophytes. Alternatively, the mussels may prevent the growth of competing algae such as *Corallina* or *Halosaccion*, allowing *Postelsia* to grow freely after wave action removes the mussels.

6

Beach Ridge

A beach ridge is a wave-swept or wave-deposited ridge running parallel to a shoreline. It is commonly composed of sand as well as sediment worked from underlying beach material. The movement of sediment by wave action is called *littoral transport*. Movement of material parallel to the shoreline is called *longshore transport*. Movement perpendicular to the shore is called *on-offshore transport*. A beach ridge may be capped by, or associated with, sand dunes. The height of a beach ridge is affected by wave size and energy.

A fall in water level (or an uplift of land) can isolate a beach ridge from the body of water that created it. Isolated beach ridges may be found along dry lakes in the western United States and inland of the Great Lakes of North America, where they formed at the end of the last ice age when lake levels were much higher due to glacial melting and obstructed outflow caused by glacial ice. Some isolated beach ridges are found in parts of Scandinavia, where glacial melting relieved pressure on land masses and resulted in subsequent crustal lifting or post-glacial rebound. A rise in water level can submerge beach ridges created at an earlier stage, causing them to erode and become less distinct. Beach ridges may become routes for roads and trails.

Beachrock

Beachrock is a friable to well-cemented sedimentary rock that consists of a variable mixture of gravel-, sand-, and silt-sized sediment that is cemented with carbonate minerals and has formed along a shoreline. Depending on location, the sediment that is cemented to form beachrock can consist of a variable mixture of shells, coral fragments, rock fragments of different types, and other materials. It can contain scattered artifacts, pieces of wood, and coconuts. Beachrock typically forms within the intertidal zone within tropical or semitropical regions. However, Quaternary beachrock is also found as far north and south as 60' latitude.

Overview

Beachrock units form under a thin cover of sediment and generally overlie unconsolidated sand. They typically consist of multiple units, representing multiple episodes of cementation and exposure. The mineralogy of beachrocks is mainly high-magnesium calcite or aragonite. The main processes involved in the cementation are : supersaturation with $CaCO_3$ through direct evaporation of seawater (Scoffin, 1970), groundwater CO_2 degassing in the vadose zone (Hanor, 1978), mixing of marine and meteoric water fluxes (Schmalz, 1971) and precipitation of micritic calcium carbonate as a byproduct of microbiological activity (Neumeier, 1999).

On retreating coasts, outcrops of beachrock may be evident offshore where they may act as a barrier against coastal erosion. Beachrock presence can also induce sediment deficiency in a beach and out-synch its wave regime. Because beachrock is lithified within the intertidal zone and because it commonly forms in a few years, its potential as an indicator of past sea level is important.

Cementation and Position of Beachrock

Beachrocks are located along the coastline in a parallel term and they are usually a few metres offshore. They are generally separated in several levels which may correspond to different generations of beachrock cementation. Thus, the older zones are located in the outer part of the formation when the younger ones are on the side of the beach, possibly under the unconsolidated sand. They also seem to have a general inclination to the sea (50 – 150). There are several appearances of beachrock formations which are characterized by multiple cracks and gaps. The result from this fact is an interruptible formation of separated blocks of beachrock, which may be of the same formation.

The length of beachrocks varies from metres to kilometers, its width can reach up to 300 metres and its height starts from 30 cm and reaches 3 metres. Following the process of coastal erosion, beachrock formation may be uncovered. Coastal erosion may be the result of sea level rise or deficit in sedimentary equilibrium. One way or another, unconsolidated sand that covers the beachrock draws away and the formation is revealed. If the process of cementation continues, new beachrock would be formed in a new position in the intertidal zone. Successive phases of sea level change may result in sequential zones of beachrock.

Pocket Beach

Pocket beach is usually a small beach, between two headlands. In an idealized setting, there is very little or no exchange of sediment between the pocket beach and the adjacent shorelines.

Pocket beaches can be natural or artificial. Many natural pocket beaches exist throughout the world. Artificial pocket beaches are usually constructed in areas where natural beaches are fairly narrow or absent. Examples of artificial pocket beaches include over 100 such systems on Chesapeake Bay in the United States, each consisting of several individual pocket beaches; the Fisher Key, Florida project constructed on a dredge spoil island originally consisting of cobble dredge material;, and the Fred Howard Park Beach that was constructed offshore of a muddy mangrove shoreline. Additionally, there have been many pocket beaches constructed in the Caribbean where resorts have been developed along rocky shorelines with minimal natural beaches.

Shingle Beach

A shingle beach is a beach which is armoured with pebbles or small- to medium-sized cobbles (as opposed to fine sand). Typically, the stone composition may grade from characteristic sizes ranging from two to 200 mm diameter. While this beach landform is most commonly found in Western Europe, examples are found in Bahrain, North America and in a number of other world regions, such as the east coast of New Zealand's South Island, where they are associated with the shingle fans of braided rivers. Though created at shorelines, post-glacial rebound can raise shingle beaches as high as 200 metres above sea level, at the High Coast in Sweden.

The ecosystems formed by this unique association of rock and sand allow colonization by a variety of rare and endangered species.

Formation

Shingle beaches are typically steep, because the waves easily flow through the coarse, porous surface of the beach, decreasing the effect of backwash erosion and increasing the formation of sediment into a steeply sloping beach.

Tourism

Shingle beaches are often criticized as undesirable for visitors. Canterbury City Council notes that the nearby shingle beach at Whitstable is uncomfortable to walk and lie on. Also, advertisers have been known to replace images of shingle beaches with sand in promotional material.

Storm Beach

***Figure:** Chesil Beach from the Isle of Portland.*

A storm beach is a beach affected by particularly fierce waves, usually with a very long fetch. The resultant landform is often a very steep beach (up to 45°) composed of rounded cobbles, shingle and occasionally sand. The stones usually have an obvious grading of pebbles, from large to small, with the larger diameter stones typically arrayed at the highest beach elevations.

Examples

A noted textbook example is the 18-mile (29 km) long Chesil Beach in Dorset, one of three major shingle structures in Britain. It is also a tombolo connecting the Isle of Portland to the mainland at Abbotsbury, west of the resort of Weymouth. Other examples appear in the Shetland and Orkney Islands, as well as the Scottish mainland at Caithness.

Wash Margin

A wash margin or wash fringe (German: *Spülsaum*) is an area of the shore on which material is deposited or washed up. It often runs along the margin of a waterbody and there can be several bands due to variations in water levels. As a result of the richness of nutrients that occur in such wash fringes, ruderal species frequently occur here, that, for example, on the Baltic Sea coast consist of grassleaf orache and sea kale.

Wind Wave

In fluid dynamics, wind waves or, more precisely, wind-generated waves are surface waves that occur on the free surface of oceans, seas, lakes, rivers, and canals or even on small puddles and ponds. They usually result from the wind blowing over a vast enough stretch of fluid surface. Waves in the oceans can travel thousands of miles before reaching land. Wind waves range in size from small ripples to huge waves over 30 m high.

When directly generated and affected by local winds, a wind wave system is called a wind sea. After the wind ceases to blow, wind waves are called *swells*. More generally, a swell consists of wind-generated waves that are not—or are hardly—affected by the local wind at that time. They have been generated elsewhere or some time ago. Wind waves in the ocean are called ocean surface waves.

Wind waves have a certain amount of randomness: subsequent waves differ in height, duration, and shape with limited predictability. They can be described as a stochastic process, in combination with the physics governing their generation, growth, propagation and decay—as well as governing the interdependence between flow quantities such as: the water surface movements, flow velocities and water pressure. The key statistics of wind waves (both seas and swells) in evolving sea states can be predicted with wind wave models.

Tsunamis are a specific type of wave not caused by wind but by geological effects. In deep water, tsunamis are not visible because they are small in height and very long in wavelength. They may grow to devastating proportions at the coast due to reduced water depth.

Wave Formation

The great majority of large breakers one observes on a beach result from distant winds. Five factors influence the formation of wind waves:

- Wind speed
- Distance of open water that the wind has blown over (called the *fetch*)
- Width of area affected by fetch
- Time duration the wind has blown over a given area
- Water depth

All of these factors work together to determine the size of wind waves. The greater each of the variables, the larger the waves. Waves are characterized by:

- Wave height (from trough to crest)
- Wave length (from crest to crest)
- Wave period (time interval between arrival of consecutive crests at a stationary point)
- Wave propagation direction

Waves in a given area typically have a range of heights. For weather reporting and for scientific analysis of wind wave statistics, their characteristic height over a period of time is usually expressed as *significant wave height*. This represents an average height of the highest one-third of the waves in a given time period (usually chosen somewhere in the range from 20 minutes to twelve hours), or in a specific wave or storm system. The significant wave height is also the value a "trained observer" (e.g. from a ship's crew) would estimate from visual observation of a sea state. Given the variability of wave height, the largest individual waves are likely to be somewhat less than twice the reported significant wave height for a particular day or storm.

Types of wind waves

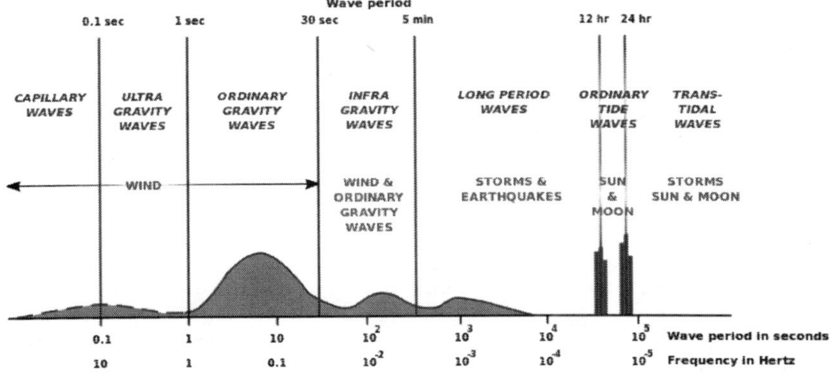

Figure: *Classification of the spectrum of ocean waves according to wave period.*

Three different types of wind waves develop over time:

- Capillary waves, or ripples
- Seas
- Swells

Ripples appear on smooth water when the wind blows, but will die quickly if the wind stops. The restoring force that allows them to propagate is surface tension. Seas are the larger-scale, often irregular motions that form under sustained winds. These waves tend to last much longer, even after the wind has died, and the restoring force that allows them to propagate is gravity. As waves propagate away from their area of origin, they naturally separate into groups of common direction and wavelength. The sets of waves formed in this way are known as swells.

Individual "rogue waves" (also called "freak waves", "monster waves", "killer waves", and "king waves") much higher than the other waves in the sea state can occur. In the case of the Draupner wave, its 25 m (82 ft) height was 2.2 times the significant wave height. Such waves are distinct from tides, caused by the Moon and Sun's gravitational pull, tsunamis that are caused by underwater earthquakes or landslides, and waves generated by underwater explosions or the fall of meteorites—all having far longer wavelengths than wind waves.

Yet, the largest ever recorded wind waves are common—not rogue—waves in extreme sea states. For example: 29.1 m (95 ft) high waves have been recorded on the RRS Discovery in a sea with 18.5 m (61 ft) significant wave height, so the highest wave is only 1.6 times the significant wave height. The biggest recorded by a buoy (as of 2011) was 32.3 m (106 ft) high during the 2007 typhoon Krosa near Taiwan.

Wave Shoaling and Refraction

As waves travel from deep to shallow water, their shape alters (wave height increases, speed decreases, and length decreases as wave orbits become asymmetrical). This process is called shoaling.

Wave refraction is the process by which wave crests realign themselves as a result of decreasing water depths. Varying depths along a wave crest cause the crest to travel at different phase speeds, with those parts of the wave in deeper water moving faster than those in shallow water. This process continues until the crests become (nearly) parallel to the depth contours. Rays—lines normal to wave crests between which a fixed amount of energy flux is contained—

converge on local shallows and shoals. Therefore, the wave energy between rays is concentrated as they converge, with a resulting increase in wave height.

Because these effects are related to a spatial variation in the phase speed, and because the phase speed also changes with the ambient current – due to the Doppler shift – the same effects of refraction and altering wave height also occur due to current variations. In the case of meeting an adverse current the wave *steepens*, i.e. its wave height increases while the wave length decreases, similar to the shoaling when the water depth decreases.

Wave Breaking

Some waves undergo a phenomenon called "breaking". A breaking wave is one whose base can no longer support its top, causing it to collapse. A wave breaks when it runs into shallow water, or when two wave systems oppose and combine forces. When the slope, or steepness ratio, of a wave is too great, breaking is inevitable.

Individual waves in deep water break when the wave steepness—the ratio of the wave height H to the wavelength λ—exceeds about 0.17, so for $H > 0.17\ \lambda$. In shallow water, with the water depth small compared to the wavelength, the individual waves break when their wave height H is larger than 0.8 times the water depth h, that is $H > 0.8\ h$. Waves can also break if the wind grows strong enough to blow the crest off the base of the wave.

Three main types of breaking waves are identified by surfers or surf lifesavers. Their varying characteristics make them more or less suitable for surfing, and present different dangers.

- Spilling, or rolling: these are the safest waves on which to surf. They can be found in most areas with relatively flat shorelines. They are the most common type of shorebreak
- Plunging, or dumping: these break suddenly and can "dump" swimmers—pushing them to the bottom with great force. These are the preferred waves for experienced surfers. Strong offshore winds and long wave periods can cause dumpers. They are often found where there is a sudden rise in the sea floor, such as a reef or sandbar.
- Surging: these may never actually break as they approach the water's edge, as the water below them is very deep. They tend to form on steep shorelines. These waves can knock swimmers over and drag them back into deeper water.

Science of Waves

Wind waves are mechanical waves that propagate along the interface between water and air; the restoring force is provided by gravity, and so they are often referred to as surface gravity waves. As the wind blows, pressure and friction forces perturb the equilibrium of the water surface. These forces transfer energy from the air to the water, forming waves. The initial formation of waves by the wind is described in the theory of Phillips from 1957, and the subsequent growth of the small waves has been modeled by Miles, also in 1957.

In the case of monochromatic linear plane waves in deep water, particles near the surface move in circular paths, making wind waves a combination of longitudinal (back and forth) and transverse (up and down) wave motions. When waves propagate in shallow water, (where the depth is less than half the wavelength) the particle trajectories are compressed into ellipses.

As the wave amplitude (height) increases, the particle paths no longer form closed orbits; rather, after the passage of each crest, particles are displaced slightly from their previous positions, a phenomenon known as Stokes drift.

As the depth below the free surface increases, the radius of the circular motion decreases. At a depth equal to half the wavelength λ, the orbital movement has decayed to less than 5% of its value at the surface. The phase speed (also called the celerity) of a surface gravity wave is – for pure periodic wave motion of small-amplitude waves – well approximated by

$$c = \sqrt{\frac{g\lambda}{2\pi}\tanh\left(\frac{2\pi d}{\lambda}\right)}$$

where

c = phase speed;

$\ddot{e}$ = wavelength;

d = water depth;

g = acceleration due to gravity at the Earth's surface.

In deep water, where $d \geq \frac{1}{2}\lambda$, so $\frac{2\pi d}{\lambda} \geq \pi$ and the hyperbolic tangent approaches 1, the speed c approximates

$$c_{\text{deep}} = \sqrt{\frac{g\lambda}{2\pi}}.$$

In SI units, with c_{deep} in m/s, $c_{\text{deep}} \approx 1.25\sqrt{\lambda}$, when $\lambda$is measured in metres. This expression tells us that waves of different wavelengths travel at different speeds. The fastest waves in a storm are the ones with the longest wavelength. As a result, after a storm, the first waves to arrive on the coast are the long-wavelength swells. For intermediate and shallow water, the Boussinesq equations are applicable, combining frequency dispersion and nonlinear effects. And in very shallow water, the shallow water equations can be used.

If the wavelength is very long compared to the water depth, the phase speed (by taking the limit of c when the wavelength approaches infinity) can be approximated by

$$c_{\text{shallow}} = \lim_{\lambda \to \infty} c = \sqrt{gd}.$$

On the other hand, for very short wavelengths, surface tension plays an important role and the phase speed of these gravity-capillary waves can (in deep water) be approximated by

$$c_{\text{gravity-capillary}} = \sqrt{\frac{g\lambda}{2\pi} + \frac{2\pi S}{\rho\lambda}}$$

where

S = surface tension of the air-water interface;

ρ= density of the water.

When several wave trains are present, as is always the case in nature, the waves form groups. In deep water the groups travel at a group velocity which is half of the phase speed. Following a single wave in a group one can see the wave appearing at the back of the group, growing and finally disappearing at the front of the group. As the water depth d decreases towards the coast, this will have an effect: wave height changes due to wave shoaling and refraction. As the wave height increases, the wave may become unstable when the crest of the wave moves faster than the trough. This causes *surf*, a breaking of the waves.

The movement of wind waves can be captured by wave energy devices. The energy density (per unit area) of regular sinusoidal waves depends on the water density ρ, gravity acceleration g and the wave height H(which, for regular waves, is equal to twice the amplitude, a):

$$E = \frac{1}{8}\rho g H^2 = \frac{1}{2}\rho g a^2.$$

The velocity of propagation of this energy is the group velocity.

Wind Wave Models

Surfers are very interested in the wave forecasts. There are many websites that provide predictions of the surf quality for the upcoming days and weeks. Wind wave models are driven by more general weather models that predict the winds and pressures over the oceans, seas and lakes.

Wind wave models are also an important part of examining the impact of shore protection and beach nourishment proposals. For many beach areas there is only patchy information about the wave climate, therefore estimating the effect of wind waves is important for managing littoral environments.

Seismic Signals

Ocean water waves generate land seismic waves that propagate hundreds of kilometers into the land. These seismic signals usually have the period of 6 ± 2 seconds. Such recordings were first reported and understood in about 1900.

There are two types of seismic "ocean waves". The primary waves are generated in shallow waters by direct water wave-land interaction and have the same period as the water waves (10 to 16 seconds).

The more powerful secondary waves are generated by the superposition of ocean waves of equal period travelling in opposite directions, thus generating standing gravity waves – with an associated pressure oscillation at half the period, which is not diminishing with depth. The theory for microseism generation by standing waves was provided by Michael Longuet-Higgins in 1950, after in 1941 Pierre Bernard suggested this relation with standing waves on the basis of observations.

Surf Zone

As ocean surface waves come closer to shore they break, forming the foamy, bubbly surface we call surf. The region of breaking waves defines the surf zone. After breaking in the surf zone, the waves (now reduced in height) continue to move in, and they run up onto the sloping front of the beach, forming an uprush of water called swash. The water then runs back again as backwash.

The nearshore zone where wave water comes onto the beach is the surf zone. The water in the breaker zone, or surf zone, is shallow, usually between 5 and 10 m (16 and 33 ft) deep; this causes the waves to be unstable.

Animal Life

The animals that often are found living in the surf zone are crabs, clams, and snails. Surf clams and mole crabs are two species that stand out as inhabitants of the surf zone. Both of these animals are very fast burrowers. The surf clam, also known as the variable coquina, is a filter feeder that uses its gills to filter microalgae, tiny zooplankton, and small particulates out of seawater. The mole crab is a suspension feeder that eats by capturing zooplankton with its antennae. All of these creatures burrow down into the sand to escape from being pulled into the ocean from the tides and waves. They also burrow themselves in the sand to protect themselves from predators. The surf zone is full of nutrients, oxygen, and sunlight which leaves the zone very productive with animal life.

Tides

The Surf Zone can help determine the danger level of rip currents. Rip Current Outlooks use the following set of qualifications:

1. Low Risk rip currents: Wind and/or wave conditions are not expected to support the development of rip currents; however, rip currents can sometimes occur, especially in the vicinity of jetties and piers. Know how to swim and heed the advice of lifeguards.
2. Moderate Risk rip currents: Wind and/or wave conditions support stronger or more frequent rip currents. Only experienced surf swimmers should enter the water.
3. High Risk rip currents: Wind and/or wave conditions support dangerous rip currents. Rip currents are life-threatening to anyone entering the surf.

Swash

Swash, in geography, is known as a turbulent layer of water that washes up on the beach after an incoming wave has broken. The swash action can move beach materials up and down the beach, which results in the cross-shore sediment exchange. The time-scale of swash motion varies from seconds to minutes depending on the type of beach. Greater swash generally occurs on flatter beaches. The swash motion plays the primary role in the formation of morphological features and their changes in the swash zone. The swash action also plays an important role as one of the instantaneous processes in wider coastal morphodynamics.

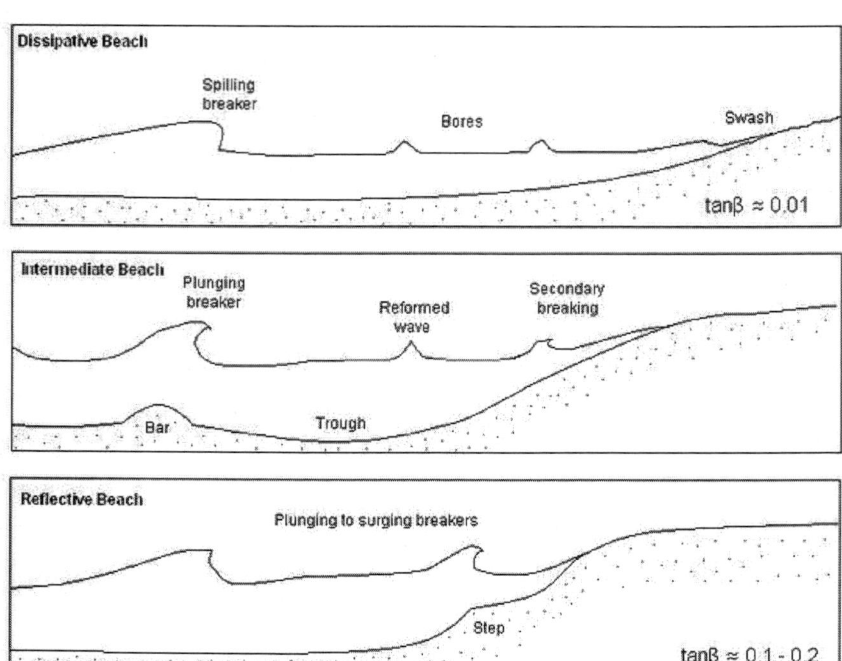

Figure: *Beach classification by Wright and Short (1983) showing dissipative, intermediate, and reflective beaches.*

There are two approaches that describe swash motions: (1) swash resulting from the collapse of high-frequency bores (f>0.05 Hz) on the beachface; and (2) swash characterised by standing, low-frequency (f<0.05 Hz) motions. Which type of swash motion prevails is dependant on the wave conditions and the beach morphology and this can be predicted by calculating the surf similarity parameter εb (Guza & Inman 1975)

$$\epsilon b = \frac{4\pi^2 Hb}{2gT^2 tan^2\beta},$$

Where Hb is the breaker height, g is gravity, T is the incident-wave period and tan β is the beach gradient. Values εb >20 indicate dissipative conditions where swash is characterised by standing long-wave motion. Values εb <2.5 indicate reflective conditions where swash is dominated by wave bores.

Contents

Uprush and Backwash: Swash consists of two phases: uprush (onshore flow) and backwash (offshore flow). Generally uprush velocities are greater but of shorter duration compared to the backwash. Onshore velocities are at greatest at the start of the uprush and then decrease, whereas offshore velocities increase towards the end of the backwash. The direction of the uprush varies with the prevailing wind, whereas the backwash is always perpendicular to the coastline. This asymmetrical motion of swash can cause longshore drift as well as cross-shore sediment transport.

Swash Morphology

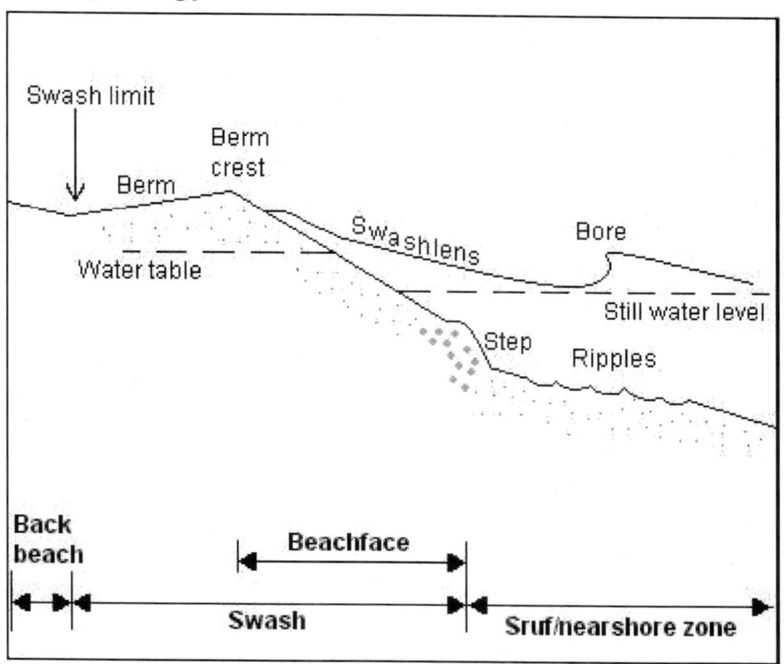

Figure: *Swash zone and beachface morphology showing terminology and principal processes (Modified from Masselink & Hughes 2003)*

The swash zone is the upper part of the beach between backbeach and surf zone, where intense erosion occurs during storms. The swash zone is alternately wet and dry. Infiltration (hydrology) and exfiltration take place between the swash flow and the beach groundwater table. Beachface, berm, beach step and beach cusps are the typical morphological features associated with swash motion. Infiltration (hydrology) and sediment transport by swash motion are important factors that govern the gradient of the beachface.

Beachface

The beachface is the planar, relatively steep section of the beach profile that is subject to swash processes . The beachface extends from the berm to the low tide level. The beachface is in dynamic equilibrium with swash action when the amount of sediment transport by uprush and backwash are equal. If the beachface is flatter than the equilibrium gradient, more sediment is transported by the uprush to result in net onshore sediment transport. If the beachface is steeper than the equilibrium gradient, the sediment transport is dominated by the backwash and this results in net offshore sediment transport. The equilibrium beachface gradient is governed by a complex interrelationship of factors such as the sediment size, permeability, and fall velocity in the swash zone as well as the wave height and the wave period. The beachface cannot be considered in isolation from the surf zone to understand the morphological changes and equilibriums as they are strongly affected by the surf zone and shoaling wave processes as well as the swash zone processes.

Berm

The berm is the relatively planar part of the swash zone where the accumulation of sediment occurs at the landward farthest of swash motion. The berm protects the backbeach and coastal dunes from waves but erosion can occur under high energy conditions such as storms. The berm is more easily defined on gravel beaches and there can be multiple berms at different elevations. On sandy beaches in contrast, the gradient of backbeach, berm and beachface can be similar. The height of the berm is governed by the maximum elevation of sediment transport during the uprush. The berm height can be predicted using the equation by Takeda and Sunamura (1982)

$$Zberm = 0.125Hb^{5/8}(gT^2)^{3/8},$$

where Hb is the breaker height, g is gravity and T is the wave period.

Beach step

The beach step is a submerged scarp at the base of the beachface. The beach steps generally comprise the coarsest material and the height can vary from several centimetres to over a metre. Beach steps form where the backwash interacts with the oncoming incident wave and generate vortex. Hughes and Cowell (1987) proposed the equation to predict the step height Zstep

$$Zstep = \sqrt{HbTws},$$

where 'ws' is the sediment fall velocity. Step height increases with increasing wave (breaker) height (Hb), wave period (T) and sediment size.

Beach Cusps

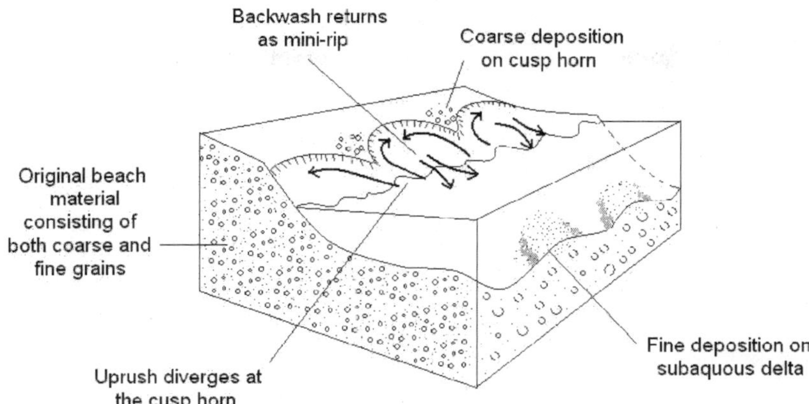

Figure: *Beach cusp morphology. Uprush diverges at the cusp horns and backwash converges in the cusp embayments. (Modified from Masselink & Hughes 2003)*

The beach cusp is a crescent-shaped accumulation of sand or gravel surrounding a semicircular depression on a beach. They are formed by swash action and more common on gravel beaches than sand. The spacing of the cusps is related to the horizontal extent of the swash motion and can range from 10 cm to 50m. Coarser sediments are found on the steep-gradient, seaward pointing 'cusp horns'. Currently there are two theories that provide an adequate explanation for the formation of the rhythmic beach cusps: standing edge waves and self-organization.

Standing Edge Wave Model

The standing edge wave theory, which was introduced by Guza and Inman (1975), suggests that swash is superimposed upon the motion of standing edge waves that travel alongshore. This produces a variation in swash height along the shore and consequently results in regular patterns of erosion. The cusp embayments form at the eroding points and cusp horns occur at the edge wave nodes. The beach cusp spacing can be predicted using the sub-harmonic edge wave model

$$\lambda = \frac{g}{\pi} T^2 tan\beta,$$

where T is incident wave period and tanâ is beach gradient.

This model only explains the initial formation of the cusps but not the continuing growth of the cusps. The amplitude of the edge wave reduces as the cusps grow, hence it is a self-limiting process.

Self-organization Model

The self-organization theory was introduced by Werner and Fink (1993) and it suggests that beach cusps form due to a combination of positive feedback that is operated by beach morphology and swash motion encouraging the topographic irregularity and negative feedback that discourages accretion or erosion on well-developed beach cusps. It is relatively recent that the computational resources and sediment transport formulations became available to show that the stable and rhythmic morphological features can be produced by such feedback systems. The beach cusp spacing, based on the self-organization model, is proportional to the horizontal extent of the swash motion S using the equation

$$\lambda = fS,$$

where the constant of proportionality f is *c.* 1.5.

Sediment Transport

Cross-shore sediment transport:

The cross-shore sediment exchange, between the subaerial and sub-aqueous zones of the beach, is primarily provided by the swash motion. The transport rates in the swash zone are much higher compared to the surf zone and suspended sediment concentrations can exceed 100 kg/m^3 close to the bed. The onshore and offshore sediment transport by swash thus plays a significant role in accretion and erosion of the beach.

There are fundamental differences in sediment transport between the uprush and backwash of the swash flow. The uprush, which is mainly dominated by bore turbulence, especially on steep beaches, generally suspend sediments to transport. Flow velocities, suspended sediment concentrations and suspended fluxes are at greatest at the start of the uprush when the turbulence is maximum. Then the turbulence dissipates towards the end of the onshore flow, settling the suspended sediment to the bed. In contrast, the backwash is dominated by the sheet flow and bedload sediment transport. The flow velocity increases towards the end of the backwash causing more bed-generated turbulence, which results in sediment transport near the bed. The direction of the net sediment transport (onshore or offshore) is largely governed by the beachface gradient.

Longshore Drift

Longshore drift by swash occurs either due to beach cusp morphology or due to oblique incoming waves causing strong alongshore swash motion. Under the influence of longshore drift, when there is no slack-water phase during backwash flows, sediments can remain suspended to result in offshore sediment transport. Beachface erosion by swash processes is not very common but erosion can occur where swash has a significant alongshore component.

Management

The swash zone is highly dynamic, accessible and susceptible to human activities. This zone can be very close to developed properties. It is said that at least 100 million people on the globe live within one metre of mean sea level. Understanding the swash zone processes and wise management is vital for coastal communities which can be affected by coastal hazards, such as erosion and storm surge.

It is important to note that the swash zone processes cannot be considered in isolation as it is strongly linked with the surf zone processes. Many other factors, including human activities and climate change, can also influence the morphodynamics in the swash zone. Understanding the wider morphodynamics is essential in successful coastal management.

Construction of sea walls has been a common tool to protect developed property, such as roads and buildings, from coastal erosion and recession. However, more often than not, protecting the property by building a seawall does not achieve the retention of the beach. Building an impermeable structure such as a seawall within the swash zone can interfere with the morphodynamics system in the swash zone. Building a seawall can raise the water table, increase wave reflection and intensify turbulence against the wall. This ultimately results in erosion of the adjacent beach or failure of the structure. Boulder ramparts (also known as revetments or riprap) and tetrapods are less reflective than impermeable sea walls, as waves are expected to break across the materials to produce swash and backwash that do not cause erosion. Rocky debris are sometimes placed in front of a sea wall in the attempt to reduce the wave impact, as well as to allow the eroded beach to recover.

Understanding the sediment transport system in the swash zone is also vital for beach nourishment projects. Swash plays a significant role in transportation and distribution of the sand that is added to the beach. There have been failures in the past due to inadequate understanding. Understanding and prediction of the sediment movements, both in the swash and surf zone, is vital for the nourishment project to succeed.

Example

The coastal management at Black Rock, on the north-east coast of Phillip Bay, Australia, provides a good example of a structural response to beach erosion which resulted in morphological changes in the swash zone. In the 1930s, a sea wall was built to protect the cliff from recession at Black Rock. This resulted in depletion of the beach in front of the sea wall, which was damaged by repeated storms in winter time. In 1969, the beach was nourished with approximately 5000m^3 of sand from inland in order to increase the volume of sand on the beach to protect the sea wall. This increased the sand volume by about 10%, however, the sand was carried away by northward drifting in autumn to leave the sea wall exposed to the impacts of winter storms again. The project had failed to take the seasonal patterns of longshore drift into account and had underestimated the amount of sand to nourish with, especially on the southern part of the beach.

Research

It is said that conduct of morphology research and field measurements in the swash zone is challenging since it is a shallow and aerated environment with rapid and unsteady swash flows. Despite the accessibility to the swash zone and the capability to take measurements with high resolution compared to the other parts of the nearshore zone, irregularity of the data has been an impediment for analysis as well as critical comparisons between theory and observation. Various and unique methods have been used for field measurements in the swash zone. For wave run-up measurements, for example, Guza and Thornton (1981, 1982) used an 80m long dual-resistance wire stretched across the beach profile and held about 3 cm above the sand by non-conducting supports.

Holman and Sallenger (1985) conducted run-up investigation by taking videos of the swash to digitise the positions of the waterline over time. Many of the studies involved engineering structures, including seawalls, jetties and breakwaters, to establish design criteria that protect the structures from overtopping by extreme run-ups. Since the 1990s, swash hydrodynamics have been more actively investigated by coastal researchers, such as Hughes M.G., Masselink J. and Puleo J.A., contributing to the better understanding of the morphodynamics in the swash zone including turbulence, flow velocities, interaction with the beach groundwater table, and sediment transport. However, the gaps in understanding still remain in swash research including turbulence, sheet flow, bedload sediment transport and hydrodynamics on ultra-dissipative beaches.

Conclusion

Swash plays an important role as one of the instantaneous coastal processes and it is as important as the long-term processes such as sea level rise and geological processes in coastal morphodynamics. Swash zone is one of the most dynamic and rapidly changing environments on the coast and it is strongly linked with the surf zone processes. Understanding the swash mechanism is essential for the understanding of formation and changes of the swash zone morphology. More importantly, understanding of the swash zone processes is vital for society to manage coast wisely. There has been significant progress in the last two decades, however, gaps in understanding and knowledge in swash research still remain today.

Sea Foam

Figure: *Sea foam generated by breaking waves*

Sea foam, ocean foam, beach foam, or spume is a type of foam created by the agitation of seawater, particularly when it contains higher concentrations of dissolved organic matter (including proteins, lignins, and lipids) derived from sources such as the offshore breakdown of algal blooms. These compounds can act as surfactants or foaming agents. As the seawater is churned by breaking waves in the surf zone adjacent to the shore, the presence of these surfactants under these turbulent conditions traps air, forming persistent bubbles that stick to each other through surface tension. Due to its low density and persistence, foam can be blown by strong on-shore winds from the beachface inland.

Hazards

Where polluted stormwater from rivers or drains discharges to the coast, sea foam formed on adjacent beaches can be polluted with viruses and other contaminants, and may have an unpleasant odour.

If crude oil discharged from tankers at sea, or motor oil, sewage and detergents from polluted stormwater are present, the resulting sea foam is even more persistent, and can have a *chocolate mousse* texture.

If the foam forms from the breakdown of a harmful algal bloom (including those caused by some dinoflagellates and cyanobacteria), direct contact with the foam, or inhalation of aerosols derived from the foam as it dries, can cause skin irritations or respiratory discomfort.

On rare occasions large amounts of sea foam up to several metres thick can accumulate at the coast and constitute a physical hazard to beach users, through concealing large rocks and voids, storm debris and, in northern New South Wales, even sea snakes.

Notable Occurrences

- A large buildup of sea foam occurred on the coast of Yamba, northern New South Wales, 29°262 S 153°212 E on 24 August 2007, and attracted world-wide media attention.
- Other sea foam occurrences at Caloundra 26°482 S 153°082 E and Point Cartwright, 26°402 47.83 S 153°082 18.13 E, on Queensland's Sunshine Coast in January–February 2008, also attracted world-wide media attention.
- During live coverage of Hurricane Irene in Ocean City, Maryland, Tucker Barnes was covered in sea foam.
- In December 2011, the coast road at Cleveleys, Lancashire was swamped by metre high drifts of sea foam.
- Following storms and high winds on 24/25 September 2012, the beach front of the Footdee area of Aberdeen was engulfed with sea foam.
- Between 27 and 28 January 2013, the Sunshine Coast in Queensland, Australia, had masses of foam wash up on land from ex-tropical cyclone Oswald.

Rip Current

A rip current, commonly referred to simply as a rip, or by the misnomer rip tide, is a strong channel of water flowing seaward from near the shore, typically through the surf line. Typical flow is at 0.5 metres per second (1–2 feet per second), and can be as fast as 2.5 metres per second (8 feet per second), which is faster than any human swimmer. They can move to different locations on a beach break, up to tens of metres (a few hundred feet) a day. They can occur at any beach with breaking waves, including oceans, seas, and large lakes.

Causes and Occurrence

When wind and waves push water towards the shore, that water is often forced sideways by the oncoming waves. This water streams along the shoreline until it finds an exit back to the sea or open lake water. The resulting rip current is usually narrow and located in a trench between sandbars, under piers or along jetties. A common misconception is that ordinary undertow or even rip currents are strong enough to pull someone *under* the surface of the water; in reality the current is strongest at the surface. This strong surface flow tends to damp incoming waves, leading to the illusion of a particularly calm part of the sea, which may possibly lure some swimmers into the area. The off-shore path taken by a rip current can be demonstrated by placing coloured dye at the start of a current at the shoreline.

Rip currents are stronger when the surf is rough (such as during high onshore winds, or when a strong hurricane is far offshore) or when the tide is low.

A more theoretical description involves a quantity known as radiation stress. This is the force (or momentum flux) exerted on the water column by the presence of the wave. As a wave shoals and increases in wave height prior to breaking, radiation stress increases. To balance this, the mean surface (the water level with the wave averaged out) decreases—this is known as *setdown*. As the wave breaks and continues to reduce in height, the radiation stress decreases. To balance this force, the mean surface increases—this is known as *setup*. As a wave propagates over a sandbar with a gap (as shown above), the wave breaks on the bar, leading to setup. However, the part of the wave that propagates over the gap does not break, and thus setdown will continue. Thus, the mean surface over the bars is higher than that over the gap, and a strong flow will issue outward through the gap.

Rip currents can potentially occur wherever strong longshore variability in wave breaking exists. This variability may be caused by sandbars (as above) or even by crossing wave trains.

Dangers

Rip currents are a source of danger for people in ocean and lake surf, dragging swimmers away from the beach. Death by drowning comes following exhaustion while fighting the current.

Although a rare event, rip currents can be deadly for non-swimmers as well: a person standing waist deep in water can be dragged into deeper waters, where they can drown if they are unable to swim and are not wearing a flotation device. Varying topography makes some beaches more likely to have rip currents; a few are notorious. Rip currents cause more than 100 deaths annually in the United States. Rip currents cause 80% of rescues needed by beach lifeguards.

Usefulness

Rip currents are often used by those who wish to leave the shoreline and want to conserve energy, such as kayakers.

Escaping a Rip Current

A swimmer caught in a rip current should not attempt to swim back to shore directly against the rip. This risks exhaustion and drowning. A rip does not pull a swimmer under water; it carries the swimmer away from the shore in a narrow channel of water. The rip is like a treadmill which the swimmer needs to step off. The swimmer should remain calm and swim parallel to the shore until he or she is outside of the current. Then, locations to aim for are places where waves are breaking. In these areas, floating objects are generally transported towards the shore.

A swimmer in a strong rip, who is unable to swim away from it, should relax and calmly float or tread water to conserve energy. Eventually the rip will lose strength, and the swimmer can swim at a leisurely pace, in a diagonal direction, away from the rip but back to shore. Coastal swimmers should understand the danger of rip currents, learn how to recognize them and how to escape from them, and swim in areas where lifeguards are operating, whenever possible.

Surf Break

A surf break (also break, shore break, or big wave break) is a permanent obstruction such as a coral reef, rock, shoal, or headland that causes a wave to break, forming a barreling wave or other wave that can be surfed, before it eventually collapses. The topography of the seabed determines the shape of the wave and type of break. Since shoals can change size and location, affecting the break, it takes commitment and skill to find good breaks. Some surf breaks are quite dangerous, since the surfer can collide with a reef or rocks below the water. Surf breaks are often defended vehemently by surfers. In 2008, surfers and environmentalists opposed a toll road project in Orange County, California that would have changed sediment patterns and affected the world-class *Trestles* surf break north of San Onofre State Beach which attracted 400,000 surfers in 2007.

In 2007, the NSW Geographical Names Register began formally recognizing names of surf breaks in Australia, defining a surf break as a "permanent obstruction such as a reef, headland, bombora, rock or sandbar, which causes waves to break".

One of the largest surf breaks in the world is the Jaws surf break in Maui, Hawaii, with waves that reach a maximum height of 40–60 feet (12–18 m).

Types of Breaks

There are several types of breaks.

Point Break

A point break refers to the place where waves hit a point of land or rocks jutting out from the coastline. Bells Beach in Australia and Jardim do Mar in Madeira, Portugal are examples of point breaks.

Beach Break

A beach break takes place where waves break on a sandy seabed. An example of a classic beach break is Hossegor in Southern France, which is famous for waves of up to 20 feet (6.1 m).

Reef Break

A reef break happens when a wave breaks over a coral reef or a rocky seabed. These waves are perhaps the most dangerous if a surfer wipes out badly. Example are Cloudbreak in Fiji and Jaws in Maui.

Shore Break

A shore break is a wave that breaks directly on, or very close to the shore. This happens when the beach is very steep at the shoreline. Oahu in Hawaii is known to have some of the largest shore breaks in the world, measuring up to a height of 3 - 4 metres during the winter months.

Coastal Hazards

Coastal Hazards are both natural and man-made disasters that happen along the coastline. This article will discuss the coastal environments on a global scale, as well as the causes of coastal hazards (ex. hurricanes and nor'easter) in which the environments are affected. Coastal policies and management and planning are later discussed to provide adequate information on implementation and mitigation of these coastal hazards.

Figure: *Cockenzie Harbour in a gale - geograph.org.uk - 370232*

== In Coastal hazards play a major role in today's society because it is a part of human nature to live near or along the coast. 80% of people live near the coast. 1.2 billion people live within 100 km of the coast and it is on the rise. It is important for us to educate ourselves and others on coastal hazards so we can continue living near the coast with the least amount of damage to the environment.

In the past, human development has effected our coastal living arrangements by making it vulnerable to such fragile environments such as the barrier islands. Disasters such as hurricanes, with high winds and swells, cause erosion along coastlines. Due to this, certain policies have been set in place to try to manage disaster property damages; FEMA, NFIP, CZM. Adaptive management has become a major source of planning in order to make development sustainable for the environment. Structural versus non-structural mitigation techniques have been a main focus for planners towards coastal hazards. Short term solutions versus long term solutions; dune, sea walls, etc., which one should planners focus on and overall which would be the better option to spend funds on. Current strategies are only an illusion of safety towards our coastlines.

For coastal hazards, it is important to have emergency management plans in advance so agencies can respond quickly and effectively. If they have better plans it will create a faster recovery time that will ensure economic, social, and environmental life back to its original state.

For example, Hurricane Katrina was a national eye opener on how on the risk of improper risk management between federal, state, and local governments. Due to the nature of the federal system and policies, often conflict with one another between both private and public agencies in addition to mandated mitigation programmes going unfunded. Communication between these governments is essential in coordinating an effective response through mitigation. Coastal hazards are unpredictable to know where and when they will occur and it is important for everyone to be prepared.

Coastal Environments

There are many different types of environments along the coasts of the United States with very diverse features that affect, influence, and mold the near-shore processes that are involved. Understanding these ecosystems and environments can further advance the mitigating techniques and policy-making efforts against natural and man-made coastal hazards in these vulnerable areas. The five most common types of coastal zones range from the northern ice-pushing, mountainous coastline of Alaska and Maine, the barrier island coasts facing the Atlantic, the steep, cliff-back headlands along the pacific coast, the marginal-sea type coastline of the Gulf region, and the coral reef coasts bordering Southern Florida and Hawaii.

Ice-pushing/Mountainous Coastline

These coastal regions along the northernmost part of the nation were affected predominantly by, along with the rest of the Pacific Coast, continuous tectonic activity, forming a very long, irregular, ridged, steep and mostly mountainous coastline. These environments are heavily occupied with permafrost and glaciers, which are the two major conditions affecting Alaska's Coastal Development.

Barrier Island Coastline

Barrier islands are a land form system that consists of fairly narrow strips of sand running parallel to the mainland and play a significant role in mitigating storm surges and oceans swells as natural storm events occur. The morphology of the various types and sizes of barrier islands depend on the wave energy, tidal range, basement controls, and sea level trends. The islands create multiple unique environments of wetland systems including marshes, estuaries, and lagoons.

Steep, Cliff-backing Abrasion Coastline

The coastline along the western part of the nation consists of very steep, cliffed rock formations generally with vegetative slopes descending down and a fringing beach below. The various sedimentary, metamorphic, and volcanic rock formations assembled along a tectonically disturbed environment, all with altering resistances running perpendicular, cause the ridged, extensive stretch of uplifted cliffs that form the peninsulas, lagoons, and valleys.

Marginal-sea Type Coastline

The southern banks of the United States border the Gulf of Mexico, intersecting numerous rivers, forming many inlets bays, and lagoons along its coast, consisting of vast areas of marsh and wetlands. This region of landform is prone to natural disasters yet highly and continuously developed, with man-made structures attaining to water flow and control.

Coral Reef Coastline

Coral reefs are located off the shores of the southern Florida and Hawaii consisting of rough and complex natural structures along the bottom of the ocean floor with extremely diverse ecosystems, absorbing up to ninety percent of the energy dissipated from wind-generated waves. This process is a significant buffer for the inner-lying coastlines, naturally protecting and minimizing the impact of storm surge and direct wave damage. Because of the highly diverse ecosystems, these coral reefs not only provide for the shoreline protection, but also deliver an abundant amount of services to fisheries and tourism, increasing its economic value.

Causes of Coastal Hazards

Natural VS Human Disasters: The population that lives along or near our coastlines are an extremely vulnerable population. There are numerous issues facing our coastlines and there are two main categories that these hazards can be placed under, National disasters and Human disasters. Both of these issues cause great damage to our coastlines and discussion is still ongoing regarding what standards or responses need to be met to help both the individuals who want to continue living along the coastline, while keeping them safe and not eroding more coastline away.

Natural disasters are disasters that are out of human control and are usually caused by the weather. Disasters that include but are not limited to; storms, tsunamis, typhoons, flooding, tides, waterspouts, nor'easters, and storm surge. Human disasters occur when humans are the main culprit behind why the disaster happened. Some human disasters are but are not limited to; pollution, trawling, and human development. Natural and human disasters continue to harm the coastlines severely and they need to be researched in order to prepare/stop the hazards if possible.

The populations that live near or along the coast experience many hazards and it affects millions of people. Around ten million people globally feel the effects of coastal problems yearly and most are due to certain natural hazards like coastal flooding with storm surges and typhoons. A major problem related to coastal regions deals with how the entire global environment is changing and in response, the coastal regions are easily effected.

Storms, Flooding, Erosion

Storms are one of the major hazards that are associated to coastal regions. Storms, flooding, and erosion are closely associated and can happen simultaneously. Tropical storms or Hurricanes especially can devastate coastal regions. For example, Florida during Hurricane Andrew occurred in 1992 that caused extreme damage. It was a category five hurricane that caused $26.5 billion in damages and even 23 individuals lost their lives from the storm. Hurricane Katrina also caused havoc along the coast to show the extreme force a hurricane can do in a certain region.

In almost all cases, storms are the major culprit that causes flooding and erosion. Flash flooding is caused by storms that occurs when a massive amount of rainfall comes down into an area over a short period of time. Where as a storm surge, which is closely related to tropical storms, is when the wind collects and pushes water towards low pressure or inland and can rise rapidly. It is an offshore rise of water and overall creates a higher sea level that rises and is pushed inland. The amount of rise or fall of storm surge depends greatly on the amount and duration of wind and water in a specific location. Also if it occurs during a high tide it can have an even greater effect on the coast.

Almost all storms with high wind and water cause erosion along the coast. Erosion occurs when but not limited to; along shore currents, tides, sea level rise and fall, and high winds. Larger amounts of erosion cause the coastline to erode away at a faster rate and can leave people homeless and leave less land to develop or keep for environmental reasons. Coastal erosion has been increasing over the past few years and it is still on the rise which makes it a major coastline hazard. In the United States, 45 percent of its coast line is along the Atlantic or Gulf coast and the erosion rate per year along the Gulf coast is at six feet a year. The average rate of erosion along the Atlantic is around two to three feet a year. Even with these findings, erosion rates in specific locations vary because of various environmental factors such as major storms that can cause major erosion upwards to 100 feet or more in only one day.

Pollution, Trawling, Human Development

Pollution, trawling, and human development are major human disasters that effect coastal regions. There are two main categories related to pollution, point source pollution, and nonpoint source pollution. Point source pollution is when there is an exact location such as a pipeline or a body of water that leads into the rivers and oceans. Known dumping into the ocean is also another point source of pollution. Nonpoint source pollution would pertain more to fertilizer runoff, and industrial waste.

Examples of pollution that effect the coastal regions are but are not limited to; fertilizer runoff, oil spills, and dumping of hazardous materials into the oceans. More human acts that hurt the coastline are as follows; waste discharge, fishing, dredging, mining, and drilling. Oil spills are one of the most hazardous dangers towards coastal communities. They are hard to contain, difficult to clean up, and devastate everything. The fish, animals such as birds, the water, and especially the coastline near the spill. The most recent oil spill that had everybody concerned with oil spill was the BP oil spill.

Trawling hurts the normal ecosystems in the water around the coastline. It depletes all ecosystems on the ocean floor such as, flounder, shellfish, marsh etc.. It is simply a giant net that is drug across the ocean floor and destroys and catches anything in its path. Human development is one of the major problems when facing coastal hazards. The overall construction of buildings and houses on the coast line takes away the natural occurrences to handle the fluctuation in water and sea level rise.

Building houses in pre-flood areas or high risk areas that are extremely vulnerable to flooding are major concerns towards human development in coastal regions. Having houses and buildings in areas that are known to have powerful storms that will create people to be in risk by living there. Also pertaining to barrier islands, where land is at risk for erosion but they still continue to build there anyway. More and more houses today are being taken by the ocean, look at picture above.

Policies

National Flood Insurance Programme: The National Flood Insurance Programme or NFIP was instituted in 1968 and offers home owners in qualifying communities an opportunity to rebuild and recover after flooding events following the decision by insurance companies to discontinue providing flood insurance. This decision was made on behalf of the private insurers after continually high and widespread flood losses. The goals of this programme are to not only better protect individuals from flood, but to reduce property losses, and reduce the total amount disbursed for flood loses by the government. Only communities which have adopted and implemented mitigation policies that are compliant with or exceed federal regulations. The regulatory policies reduce risk to life and property located within floodplains. The NFIP also comprehensively mapped domestic floodplains increasing public awareness of risk. The majority of structures were constructed after the mapping was completed and risk could be assessed. To reduce the cost to these owners, which consititue roughly 25% of the total policies the rates for insurance are subsidized.

Coastal States Organization

The Coastal States Organization or CSO was established in 1970 to represent 35 U.S. sub-federal governments on issues of coastal policies. CSO lobbies Congress on issues pertaining to Coastal Policy allowing states input on federal policy decisions. Funding, support, water quality, coastal hazards, and coastal zone management are the primary issues CSO promotes. The strategic goals of CSO are to provide information and assistance to members,evaluate and manage coastal needs, and secure long term funding for member states initiatives.

Coastal Zone Management Act

In 1972 the Coastal Zone Management Act or CZMA works to streamline the policies which states create to a minimum federal standard for environmental protection. CZMA establishes the national policy for the development and implementation of regulatory programmes for coastal land usage, which is supposed to be reflected in state legislation such as CAMA. CZMA also provides minimum building requirements to make the insurance provided through the NFIP less expensive for the government to operate by mitigating losses. Congress found that it was necessary to establish the minimum which programmes should provide for. Each coastal state is required to have a programme with 7 distinct parts: Identifying land uses, Identifying critical coastal areas, Management measures, Technical assistance, Public participation, Administrative coordination, State coastal zone boundary modification.

The Coastal Area Management Act

The Coastal Area Management Act or CAMA is policy that was implemented by the state of North Carolina in 1974 to work in-tandem with the CZMA. It creates a cooperative programme between the state and local governments. The State government operates in an advisory capacity and reviews decisions made by local government planners. The goal of this legislation was to create a management system capable of preserving the coastal environment, insure the preservation of land and water resources, balance the use of coastal resources and establish guidelines and standards for conservations, economic development, tourism, transportation, and the protection of common law.

Management and Planning

Due to the increasing urbanization along the coastlines, planning and management are essential to protecting the ecosystems and environment from depleting. Coastal management is becoming implemented more because of the movement of people to the shore and the hazards that come with the territory. Some of the hazards include movement of barrier islands, sea level rise, hurricanes, nor'easters, earthquakes, flooding, erosion, pollution and human development along the coast.

The Coastal Zone Management Act (CZMA) was created in 1972 because of the continued growth along the coast, this act introduced better management practices such as integrated coastal zone management, adaptive management and the use mitigation strategies when planning. According to the Coastal Zone Management Act, the objectives are to remain balanced to "preserve, protect, develop, and where possible, to restore or enhance the resources of the nation's coastal zone".

The development of the land can strongly affect the sea, for example the engineering of structures versus non-structures and the effects of erosion along the shore.

Integrated Coastal Zone Management

Integrated coastal zone management means the integration of all aspects of the coastal zone; this includes environmentally, socially, culturally politically and economically to meet a sustainable balance all around. Sustainability is the goal to allow development yet protect the environment in which we develop. Coastal zones are fragile and do not do well with change so it is important to acquire sustainable development. The integration from all views will entitle a holistic view for the best implementation and management of that country, region and local scales.

The five types of integration include integration among sectors, integration between land and water elements of the coastal zone, integration amount levels of government, integration between nations and integration among disciplines are all essential to meet the needs for implementation. Management practices include

1. maintaining the functional integrity of the coastal resource systems, without disrupting the environment
2. reducing resource-use conflicts, by making sure resources are used adequately and sustainably,
3. maintaining the health of the environment, which means to protect the ecosystems and natural cycle,
4. facilitating the progress of multisectoral development, which means allowing developers to develop within standards.

These four management practices should be based on a bottom-up approach, meaning the approach starts from a local level which is more intimate to the specific environment of that area. After assessment from the local level, the state and federal input can be implemented.

The bottom-up approach is key for protecting the local environments because there is a diversity of environments that have specific needs all over the world.

Adaptive Management

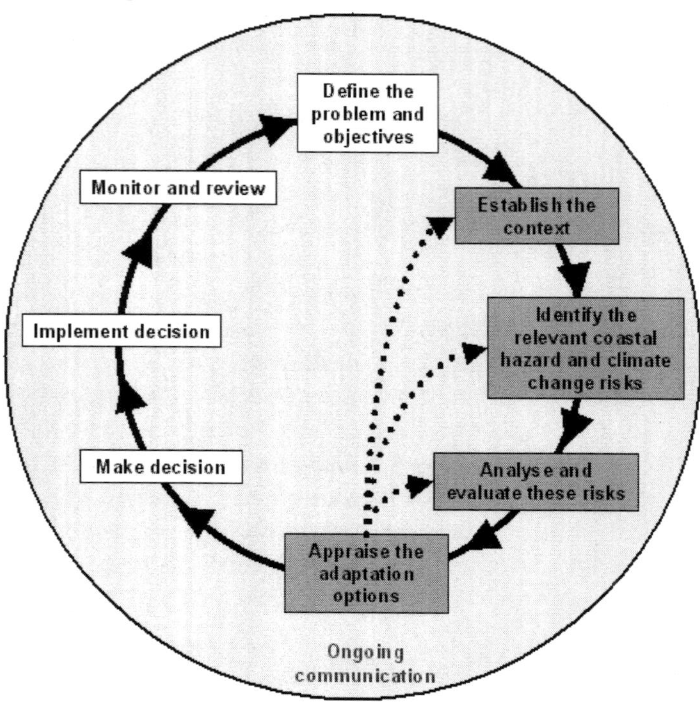

Figure: *Managing Coastal Hazards Chart*

Adaptive management is another practice of development adaptation with the environment. Resources are the major factor when managing adaptively to a certain environment to accommodate all the needs of development and ecosystems. Strategies used must be flexible by either passive or active adaptive management include these key features:

- AIterative decision-making (evaluating results and adjusting actions on the basis of what has been learned)
- Feedback between monitoring and decisions (learning process)
- Explicit characterization of system uncertainty through multi-model inference (experimentation)
- Embracing risk and uncertainty as a way of building understanding (trial and error)

To achieve adaptive management is testing the assumptions to achieve a desired outcome, such as trial and error, find the best known strategy then monitoring it to adapt to the environment, and learning the outcomes of success and failures of a project.

Mitigation

The purpose of mitigation is not only to minimize the loss of property damage, but minimize environmental damages due to development. To avoid impacts by not taking or limiting actions, to reduce or rectify impacts by rehabilitation or restoring the affected environments or instituting long-term maintenance operations and compensating for impacts by replacing or providing substitute environments for resources Structural mitigation is the current solution to eroding beaches and movement of sand is the use of engineered structures along the coast have been short lived and are only an illusion of safety to the public that result in long term damage of the coastline. Structural management deals with the use of the following: groins which are man-made solution to longshore current movements up and down the coast.

The use of groins are efficient to some extent yet cause erosion and sand build up father down the beaches. Bulkheads are man-made structures that help protect the homes built along the coast and other bodies of water that actually induce erosion in the long run. Jetties are structures built to protect sand movement into the inlets where boats for fishing and recreation move through. The use of nonstructural mitigation is the practice of using organic and soft structures for solutions to protect against coastal hazards. These include: artificial dunes, which are used to create dunes that have been either developed on or eroded. There needs to be at least two lines of dunes before any development can occur. Beach Nourishment is a major source of nonstructural mitigation to ensure that beaches are present for the communities and for the protection of the coastline. Vegetation is a key factor when protecting from erosion, specifically for to help stabilize dune erosion.

7

Marine Pollution

Marine pollution occurs when harmful, or potentially harmful, effects result from the entry into the ocean of chemicals, particles, industrial, agricultural and residential waste, noise, or the spread of invasive organisms. Most sources of marine pollution are land based. The pollution often comes from nonpoint sources such as agricultural runoff and wind blown debris and dust. Nutrient pollution, a form of water pollution, refers to contamination by excessive inputs of nutrients. It is a primary cause of eutrophication of surface waters, in which excess nutrients, usually nitrogen or phosphorus, stimulate algal growth.

Many potentially toxic chemicals adhere to tiny particles which are then taken up by plankton and benthos animals, most of which are either deposit or filter feeders. In this way, the toxins are concentrated upward within ocean food chains. Many particles combine chemically in a manner highly depletive of oxygen, causing estuaries to become anoxic.

When pesticides are incorporated into the marine ecosystem, they quickly become absorbed into marine food webs. Once in the food webs, these pesticides can cause mutations, as well as diseases, which can be harmful to humans as well as the entire food web.

Toxic metals can also be introduced into marine food webs. These can cause a change to tissue matter, biochemistry, behaviour, reproduction, and suppress growth in marine life. Also, many animal feeds have a high fish meal or fish hydrolysate content. In this way, marine toxins can be transferred to land animals, and appear later in meat and dairy products.

History

Although marine pollution has a long history, significant international laws to counter it were only enacted in the twentieth century. Marine pollution was a concern during several United Nations Conferences on the Law of the Sea beginning in the 1950s. Most scientists believed that the oceans were so vast that they had unlimited ability to dilute, and thus render pollution, harmless.

In the late 1950s and early 1960s, there were several controversies about dumping radioactive waste off the coasts of the United States by companies licensed by the Atomic Energy Commission, into the Irish Sea from the British reprocessing facility at Windscale, and into the Mediterranean Sea by the French Commissariat à l'Energie Atomique. After the Mediterranean Sea controversy, for example, Jacques Cousteau became a worldwide figure in the campaign to stop marine pollution. Marine pollution made further international headlines after the 1967 crash of the oil tanker Torrey Canyon, and after the 1969 Santa Barbara oil spill off the coast of California.

Marine pollution was a major area of discussion during the 1972 United Nations Conference on the Human Environment, held in Stockholm. That year also saw the signing of the Convention on the Prevention of Marine Pollution by Dumping of Wastes and Other Matter, sometimes called the London Convention. The London Convention did not ban marine pollution, but it established black and gray lists for substances to be banned (black) or regulated by national authorities (gray). Cyanide and high-level radioactive waste, for example, were put on the black list. The London Convention applied only to waste dumped from ships, and thus did nothing to regulate waste discharged as liquids from pipelines.

Pathways of Pollution

There are many different ways to categorize, and examine the inputs of pollution into our marine ecosystems. Patin (n.d.) notes that generally there are three main types of inputs of pollution into the ocean: direct discharge of waste into the oceans, runoff into the waters due to rain, and pollutants that are released from the atmosphere.

One common path of entry by contaminants to the sea are rivers. The evaporation of water from oceans exceeds precipitation. The balance is restored by rain over the continents entering rivers and then being returned to the sea. The Hudson in New York State and the Raritan in New Jersey, which empty at the northern and southern ends of Staten Island, are a source of mercury contamination of zooplankton (copepods) in the open ocean. The highest concentration in the filter-feeding copepods is not at the mouths of these rivers but 70 miles south, nearer Atlantic City, because water flows close to the coast. It takes a few days before toxins are taken up by the plankton.

Pollution is often classed as point source or nonpoint source pollution. Point source pollution occurs when there is a single, identifiable, and localized source of the pollution. An example is directly discharging sewage and industrial waste into the ocean. Pollution such as this occurs particularly in developing nations. Nonpoint source pollution occurs when the pollution comes from ill-defined and diffuse sources. These can be difficult to regulate. Agricultural runoff and wind blown debris are prime examples.

Direct Discharge

Pollutants enter rivers and the sea directly from urban sewerage and industrial waste discharges, sometimes in the form of hazardous and toxic wastes.

Inland mining for copper, gold. etc., is another source of marine pollution. Most of the pollution is simply soil, which ends up in rivers flowing to the sea. However, some minerals discharged in the course of the mining can cause problems, such as copper, a common industrial pollutant, which can interfere with the life history and development of coral polyps. Mining has a poor environmental track record. For example, according to the United States Environmental Protection Agency, mining has contaminated portions of the headwaters of over 40% of watersheds in the western continental US. Much of this pollution finishes up in the sea.

Land Runoff

Surface runoff from farming, as well as urban runoff and runoff from the construction of roads, buildings, ports, channels, and harbours, can carry soil and particles laden with carbon, nitrogen, phosphorus, and minerals. This nutrient-rich water can cause fleshy algae and phytoplankton to thrive in coastal areas; known as algal blooms, which have the potential to create hypoxic conditions by using all available oxygen.

Polluted runoff from roads and highways can be a significant source of water pollution in coastal areas. About 75 percent of the toxic chemicals that flow into Puget Sound are carried by stormwater that runs off paved roads and driveways, rooftops, yards and other developed land.

Ship Pollution

Ships can pollute waterways and oceans in many ways. Oil spills can have devastating effects. While being toxic to marine life, polycyclic aromatic hydrocarbons (PAHs), found in crude oil, are very difficult to clean up, and last for years in the sediment and marine environment.

Discharge of cargo residues from bulk carriers can pollute ports, waterways and oceans. In many instances vessels intentionally discharge illegal wastes despite foreign and domestic regulation prohibiting such actions. It has been estimated that container ships lose over 10,000 containers at sea each year (usually during storms). Ships also create noise pollution that disturbs natural wildlife, and water from ballast tanks can spread harmful algae and other invasive species.

Ballast water taken up at sea and released in port is a major source of unwanted exotic marine life. The invasive freshwater zebra mussels, native to the Black, Caspian and Azov seas, were probably transported to the Great Lakes via ballast water from a transoceanic vessel. Meinesz believes that one of the worst cases of a single invasive species causing harm to an ecosystem can be attributed to a seemingly harmless jellyfish. *Mnemiopsis leidyi*, a species of comb jellyfish that spread so it now inhabits estuaries in many parts of the world.

It was first introduced in 1982, and thought to have been transported to the Black Sea in a ship's ballast water. The population of the jellyfish shot up exponentially and, by 1988, it was wreaking havoc upon the local fishing industry.

> *"The anchovy catch fell from 204,000 tons in 1984 to 200 tons in 1993; sprat from 24,600 tons in 1984 to 12,000 tons in 1993; horse mackerel from 4,000 tons in 1984 to zero in 1993."*

Now that the jellyfish have exhausted the zooplankton, including fish larvae, their numbers have fallen dramatically, yet they continue to maintain a stranglehold on the ecosystem.

Invasive species can take over once occupied areas, facilitate the spread of new diseases, introduce new genetic material, alter underwater seascapes and jeopardize the ability of native species to obtain food. Invasive species are responsible for about $138 billion annually in lost revenue and management costs in the US alone.

Atmospheric Pollution

Another pathway of pollution occurs through the atmosphere. Wind blown dust and debris, including plastic bags, are blown seaward from landfills and other areas. Dust from the Sahara moving around the southern periphery of the subtropical ridge moves into the Caribbean and Florida during the warm season as the ridge builds and moves northward through the subtropical Atlantic. Dust can also be attributed to a global transport from the Gobi and Taklamakan deserts across Korea, Japan, and the Northern Pacific to the Hawaiian Islands. Since 1970, dust outbreaks have worsened due to periods of drought in Africa.

There is a large variability in dust transport to the Caribbean and Florida from year to year; however, the flux is greater during positive phases of the North Atlantic Oscillation. The USGS links dust events to a decline in the health of coral reefs across the Caribbean and Florida, primarily since the 1970s.

Climate change is raising ocean temperatures and raising levels of carbon dioxide in the atmosphere. These rising levels of carbon dioxide are acidifying the oceans. This, in turn, is altering aquatic ecosystems and modifying fish distributions, with impacts on the sustainability of fisheries and the livelihoods of the communities that depend on them. Healthy ocean ecosystems are also important for the mitigation of climate change.

Deep Sea Mining

Deep sea mining is a relatively new mineral retrieval process that takes place on the ocean floor. Ocean mining sites are usually around large areas of polymetallic nodules or active and extinct hydrothermal vents at about 1,400 - 3,700 metres below the ocean's surface. The vents create sulfide deposits, which contain precious metals such as silver, gold, copper, manganese, cobalt, and zinc. The deposits are mined using either hydraulic pumps or bucket systems that take ore to the surface to be processed. As with all mining operations, deep sea mining raises questions about environmental damages to the surrounding areas

Because deep sea mining is a relatively new field, the complete consequences of full scale mining operations are unknown. However, experts are certain that removal of parts of the sea floor will result in disturbances to the benthic layer, increased toxicity of the water column and sediment plumes from tailings. Removing parts of the sea floor disturbs the habitat of benthic organisms, possibly, depending on the type of mining and location, causing permanent disturbances. Aside from direct impact of mining the area, leakage, spills and corrosion would alter the mining area's chemical makeup.

Among the impacts of deep sea mining, sediment plumes could have the greatest impact. Plumes are caused when the tailings from mining (usually fine particles) are dumped back into the ocean, creating a cloud of particles floating in the water. Two types of plumes occur: near bottom plumes and surface plumes. Near bottom plumes occur when the tailings are pumped back down to the mining site.

The floating particles increase the turbidity, or cloudiness, of the water, clogging filter-feeding apparatuses used by benthic organisms. Surface plumes cause a more serious problem. Depending on the size of the particles and water currents the plumes could spread over vast areas. The plumes could impact zooplankton and light penetration, in turn affecting the food web of the area.

Types of Pollution

Acidification: The oceans are normally a natural carbon sink, absorbing carbon dioxide from the atmosphere. Because the levels of atmospheric carbon dioxide are increasing, the oceans are becoming more acidic. The potential consequences of ocean acidification are not fully understood, but there are concerns that structures made of calcium carbonate may become vulnerable to dissolution, affecting corals and the ability of shellfish to form shells.

Oceans and coastal ecosystems play an important role in the global carbon cycle and have removed about 25% of the carbon dioxide emitted by human activities between 2000 and 2007 and about half the anthropogenic CO_2 released since the start of the industrial revolution. Rising ocean temperatures and ocean acidification means that the capacity of the ocean carbon sink will gradually get weaker, giving rise to global concerns expressed in the Monaco and Manado Declarations.

A report from NOAA scientists published in the journal Science in May 2008 found that large amounts of relatively acidified water are upwelling to within four miles of the Pacific continental shelf area of North America. This area is a critical zone where most local marine life lives or is born. While the paper dealt only with areas from

Vancouver to northern California, other continental shelf areas may be experiencing similar effects.

A related issue is the methane clathrate reservoirs found under sediments on the ocean floors. These trap large amounts of the greenhouse gas methane, which ocean warming has the potential to release. In 2004 the global inventory of ocean methane clathrates was estimated to occupy between one and five million cubic kilometres. If all these clathrates were to be spread uniformly across the ocean floor, this would translate to a thickness between three and fourteen metres. This estimate corresponds to 500-2500 gigatonnes carbon (Gt C), and can be compared with the 5000 Gt C estimated for all other fossil fuel reserves.

Eutrophication

Eutrophication is an increase in chemical nutrients, typically compounds containing nitrogen or phosphorus, in an ecosystem. It can result in an increase in the ecosystem's primary productivity (excessive plant growth and decay), and further effects including lack of oxygen and severe reductions in water quality, fish, and other animal populations.

The biggest culprit are rivers that empty into the ocean, and with it the many chemicals used as fertilizers in agriculture as well as waste from livestock and humans. An excess of oxygen depleting chemicals in the water can lead to hypoxia and the creation of a dead zone.

Estuaries tend to be naturally eutrophic because land-derived nutrients are concentrated where runoff enters the marine environment in a confined channel. The World Resources Institute has identified 375 hypoxic coastal zones around the world, concentrated in coastal areas in Western Europe, the Eastern and Southern coasts of the US, and East Asia, particularly in Japan. In the ocean, there are frequent red tide algae blooms that kill fish and marine mammals and cause respiratory problems in humans and some domestic animals when the blooms reach close to shore.

In addition to land runoff, atmospheric anthropogenic fixed nitrogen can enter the open ocean. A study in 2008 found that this could account for around one third of the ocean's external (non-recycled) nitrogen supply and up to three per cent of the annual new marine biological production. It has been suggested that accumulating reactive nitrogen in the environment may have consequences as serious as putting carbon dioxide in the atmosphere.

One proposed solution to eutrophication in estuaries is to restore shellfish populations, such as oysters. Oyster reefs remove nitrogen from the water column and filter out suspended solids, subsequently reducing the likelihood or extent of harmful algal blooms or anoxic conditions. Filter feeding activity is considered beneficial to water quality by controlling phytoplanton density and sequestering nutrients, which can be removed from the system through shellfish harvest, buried in the sediments, or lost through denitrification. Foundational work towards the idea of improving marine water quality through shellfish cultivation to was conducted by Odd Lindahl et al., using mussels in Sweden.

Plastic Debris

Marine debris is mainly discarded human rubbish which floats on, or is suspended in the ocean. Eighty percent of marine debris is plastic - a component that has been rapidly accumulating since the end of World War II. The mass of plastic in the oceans may be as high as one hundred million metric tons.

Discarded plastic bags, six pack rings and other forms of plastic waste which finish up in the ocean present dangers to wildlife and fisheries. Aquatic life can be threatened through entanglement, suffocation, and ingestion. Fishing nets, usually made of plastic, can be left or lost in the ocean by fishermen. Known as ghost nets, these entangle fish, dolphins, sea turtles, sharks, dugongs, crocodiles, seabirds, crabs, and other creatures, restricting movement, causing starvation, laceration and infection, and, in those that need to return to the surface to breathe, suffocation.

Many animals that live on or in the sea consume flotsam by mistake, as it often looks similar to their natural prey. Plastic debris, when bulky or tangled, is difficult to pass, and may become permanently lodged in the digestive tracts of these animals, blocking the passage of food and causing death through starvation or infection.

Plastics accumulate because they don't biodegrade in the way many other substances do. They will photodegrade on exposure to the sun, but they do so properly only under dry conditions, and water inhibits this process. In marine environments, photodegraded plastic disintegrates into ever smaller pieces while remaining polymers, even down to the molecular level. When floating plastic particles photodegrade down to zooplankton sizes, jellyfish attempt to consume them, and in this way the plastic enters the ocean food chain. Many of these long-lasting pieces end up in the stomachs of marine birds and animals, including sea turtles, and black-footed albatross.

Plastic debris tends to accumulate at the centre of ocean gyres. In particular, the Great Pacific Garbage Patch has a very high level of plastic particulate suspended in the upper water column. In samples taken in 1999, the mass of plastic exceeded that of zooplankton (the dominant animal life in the area) by a factor of six. Midway Atoll, in common with all the Hawaiian Islands, receives substantial amounts of debris from the garbage patch. Ninety percent plastic, this debris accumulates on the beaches of Midway where it becomes a hazard to the bird population of the island. Midway Atoll is home to two-thirds (1.5 million) of the global population of Laysan Albatross. Nearly all of these albatross have plastic in their digestive system and one-third of their chicks die.

Toxic additives used in the manufacture of plastic materials can leach out into their surroundings when exposed to water. Waterborne hydrophobic pollutants collect and magnify on the surface of plastic debris, thus making plastic far more deadly in the ocean than it would be on land. Hydrophobic contaminants are also known to bioaccumulate in fatty tissues, biomagnifying up the food chain and putting pressure on apex predators. Some plastic additives are known to disrupt the endocrine system when consumed, others can suppress the immune system or decrease reproductive rates. Floating debris can also absorb persistent organic pollutants from seawater, including PCBs, DDT and PAHs. Aside from toxic effects, when ingested some of these are mistaken by the animal brain for estradiol, causing hormone disruption in the affected wildlife.

Toxins

Apart from plastics, there are particular problems with other toxins that do not disintegrate rapidly in the marine environment. Examples of persistent toxins are PCBs, DDT, pesticides, furans, dioxins, phenols and radioactive waste. Heavy metals are metallic chemical elements that have a relatively high density and are toxic or poisonous at low concentrations. Examples are mercury, lead, nickel, arsenic and cadmium. Such toxins can accumulate in the tissues of many species of aquatic life in a process called bioaccumulation. They are also known to accumulate in benthic environments, such as estuaries and bay muds: a geological record of human activities of the last century.

Specific examples

- Chinese and Russian industrial pollution such as phenols and heavy metals in the Amur River have devastated fish stocks and damaged its estuary soil.
- Wabamun Lake in Alberta, Canada, once the best whitefish lake in the area, now has unacceptable levels of heavy metals in its sediment and fish.
- Acute and chronic pollution events have been shown to impact southern California kelp forests, though the intensity of the impact seems to depend on both the nature of the contaminants and duration of exposure.
- Due to their high position in the food chain and the subsequent accumulation of heavy metals from their diet, mercury levels can be high in larger species such as bluefin and albacore. As a result, in March 2004 the United States FDA issued guidelines recommending that pregnant women, nursing mothers and children limit their intake of tuna and other types of predatory fish.
- Some shellfish and crabs can survive polluted environments, accumulating heavy metals or toxins in their tissues. For example, mitten crabs have a remarkable ability to survive in highly modified aquatic habitats, including polluted waters. The farming and harvesting of such species needs careful management if they are to be used as a food.
- Surface runoff of pesticides can alter the gender of fish species genetically, transforming male into female fish.
- Heavy metals enter the environment through oil spills - such as the Prestige oil spill on the Galician coast - or from other natural or anthropogenic sources.
- In 2005, the 'Ndrangheta, an Italian mafia syndicate, was accused of sinking at least 30 ships loaded with toxic waste, much of it radioactive. This has led to widespread investigations into radioactive-waste disposal rackets.
- Since the end of World War II, various nations, including the Soviet Union, the United Kingdom, the United States, and Germany, have disposed of chemical weapons in the Baltic Sea, raising concerns of environmental contamination.

Underwater Noise

Marine life can be susceptible to noise or sound pollution from sources such as passing ships, oil exploration seismic surveys, and naval low-frequency active sonar. Sound travels more rapidly and over larger distances in the sea than in the atmosphere. Marine animals, such as cetaceans, often have weak eyesight, and live in a world largely defined by acoustic information. This applies also to many deeper sea fish, who live in a world of darkness. Between 1950 and 1975, ambient noise in the ocean increased by about ten decibels (that is a tenfold increase).

Noise also makes species communicate louder, which is called the Lombard vocal response. Whale songs are longer when submarine-detectors are on. If creatures don't "speak" loud enough, their voice can be masked by anthropogenic sounds. These unheard voices might be warnings, finding of prey, or preparations of net-bubbling. When one species begins speaking louder, it will mask other species voices, causing the whole ecosystem to eventually speak louder.

According to the oceanographer Sylvia Earle, "Undersea noise pollution is like the death of a thousand cuts. Each sound in itself may not be a matter of critical concern, but taken all together, the noise from shipping, seismic surveys, and military activity is creating a totally different environment than existed even 50 years ago. That high level of noise is bound to have a hard, sweeping impact on life in the sea."

Adaptation and Mitigation

Much anthropogenic pollution ends up in the ocean. The 2011 edition of the United Nations Environment Programme Year Book identifies as the main emerging environmental issues the loss to the oceans of massive amounts of phosphorus, "a valuable fertilizer needed to feed a growing global population", and the impact billions of pieces of plastic waste are having globally on the health of marine environments. Bjorn Jennssen (2003) notes in his article, "Anthropogenic pollution may reduce biodiversity and productivity of marine ecosystems, resulting in reduction and depletion of human marine food resources". There are two ways the overall level of this pollution can be mitigated: either the human population is reduced, or a way is found to reduce the ecological footprint left behind by the average human. If the second way is not adopted, then the first way may be imposed as world ecosystems falter.

The second way is for humans, individually, to pollute less. That requires social and political will, together with a shift in awareness so more people respect the environment and are less disposed to abuse it. At an operational level, regulations, and international government participation is needed. It is often very difficult to regulate marine pollution because pollution spreads over international barriers, thus making regulations hard to create as well as enforce.

Without appropriate awareness of marine pollution, the necessary global will to effectively address the issues may prove inadequate. Balanced information on the sources and harmful effects of marine pollution need to become part of general public awareness, and ongoing research is required to fully establish, and keep current, the scope of the issues. As expressed in Daoji and Dag's research, one of the reasons why environmental concern is lacking among the Chinese is because the public awareness is low and therefore should be targeted. Likewise, regulation, based upon such in-depth research should be employed.

In California, such regulations have already been put in place to protect Californian coastal waters from agricultural runoff. This includes the California Water Code, as well as several voluntary programmes. Similarly, in India, several tactics have been employed that help reduce marine pollution, however, they do not significantly target the problem. In Chennai, sewage has been dumped further into open waters. Due to the mass of waste being deposited, open-ocean is best for diluting, and dispersing pollutants, thus making them less harmful to marine ecosystems.

Beach Hut

A beach hut (also known as a beach cabin or bathing box) is a small, usually wooden and often brightly coloured, box above the high tide mark on popular bathing beaches. They are generally used as a shelter from the sun or wind, changing into and out of swimming costumes and for the safe storing of some personal belongings. Some beach huts incorporate simple facilities for preparing food and hot drinks by either bottled gas or occasionally mains electricity.

Locations

At many seaside resorts, beach huts are arranged in one or more ranks along the top of the beach. Depending upon the location, beach huts may be owned privately or may be owned by the local council or similar administrative body. On popular beaches, privately owned beach huts can command substantial prices due to their convenient location, out of all proportion to their size and amenity. A pre-war wooden beach chalet at West Bexington, Dorset sold at auction for £216,000 in 2006, and a beach hut on Mudeford Spit sold for £170,000 in 2012. However these were exceptional as in both cases overnight stays were possible. Prices in 2009 for typical huts around the UK started from £6,000 in Walton on the Naze and typically up to £35,000.

Today there are believed to be around 20,000 beach huts in the U.K. Locations where beach huts can be seen include Lowestoft, Southwold, Walton-on-the-Naze, Abersoch, Langland Bay, Rotherslade, Rustington, St Helens, Isle of Wight, and Mersea Island. Locations in other countries include Wimereux, France, Cape Town, South Africa and Brighton and elsewhere around Port Phillip, Australia.

History

Figure: *Beach huts at Great Yarmouth, Norfolk*

The noted bathing boxes at Brighton in Australia are known to have existed as far back as 1862. The bathing boxes are thought to have been constructed and used largely as a response to the Victorian morality of the age, and are known to have existed not only in Australia but also on the beaches of England, France and Italy at around the same time.

They had evolved from the wheeled bathing machines used by Victorians to preserve their modesty. George III used a bathing machine at Mudeford in 1801, while Queen Victoria installed one at Osbourne House on the Isle of Wight in the 1840s.

In the early 20th century, beach huts were regarded as "holiday homes for the toiling classes", but in the 1930s their image revived, George V and Queen Mary spent the day at a beach hut in Sussex, and other owners have included the Spencer family and Laurence Olivier. During World War II all UK beaches were closed, the reopening in the late 1940s and 1950s led to resurgence of the British beach holiday and the heyday of the Beach Hut.

While many beach huts were former fishermen's huts, boatsheds or converted bathing machines some of the earliest purpose built beach huts in the UK were erected at Bournemouth, either side of Bournemouth Pier in 1909. Designed by F. P. Dolamore, Bournemouth's Borough Engineer, they were offered for hire for £12 10s per year. 160 huts, or bungalows as they were styled, were initially built before the first world war. The Council believes that some of the original huts are still standing, including the oldest hut number 2359 which is marked with a blue plaque, although the majority were renewed in the 1950s and 1960s. Today, Bournemouth features around 520 huts owned by the Council, There are in addition a further 1200 privately owned huts. They vary in style from the traditional, wooden, shed-like constructions to the ultra modern, concrete terrace style huts such as the 1950s Overstrand beach huts at Boscombe. These were renovated to designs by Wayne and Gerardine Hemingway, founders of the Red or Dead label, as Beach Pods for the Surf Reef opened in Autumn 2009.

Notable Huts

The Queen's beach hut in Norfolk was destroyed by fire in 2003. The building had been owned by the Royal Family for 70 years and was known to be much loved by the Queen. More recently the artist, Tracey Emin, sold her Whitstable beach hut to the collector, Charles Saatchi, for £75,000. This hut was also destroyed by fire when the warehouse where it was stored burnt down. In April 2011 Bournemouth Council obtained planning permission to site a beach hut "chapel" on the sand to host wedding and civil partnership ceremonies. The "super beach hut" is located on Bournemouth's beach under the West Cliff lift.

Beachcombing

Beachcombing is an activity that consists of an individual "combing" (or searching) the beach and the intertidal zone, looking for things of value, interest or utility. A beachcomber is a person who participates in the activity of beachcombing. Despite these general definitions, beachcombing and beachcomber are words with multiple, but related, meanings that have evolved over time.

Historical Usage

The first appearance of the word "beachcombers" in print was in Herman Melville's *Omoo* (1847). It described a population of Europeans who lived in South Pacific islands, "combing" the beach and nearby water for flotsam, jetsam, or anything else they could use or trade. When a beachcomber became totally dependent upon coastal fishing for his sustenance, or abandoned his original culture and set of values, then the term "beachcomber" was synonymous with a criminal, a drifter, or a bum.

The vast majority of beachcombers however, were simply unemployed sailors like Herman Melville in *Typee*, or Harry Franck in the book *Vagabonding Around the World.*

After enduring a voyage of danger and hardship, it was not uncommon for a few whalemen to desert a ship when it arrived in Tahiti or the Marquesas and reside, at least for a while, in the South Sea islands of Polynesia. If another beachcomber was ready to take his place in order to get home, the captain might let the disgruntled crewman go; otherwise, the captain would offer the natives a reward to find and return the deserter, and deduct the reward, plus interest, from the deserter's pay.

In other words, the deserter, if caught, would end up working the entire voyage for no pay at all, or even return home in debt to his employers. In *Typee*, Melville deserted, not once but twice, before signing on as a crewman on a Navy frigate, without fear of repercussions.

Some beachcombers traded between local tribes, and between tribes and visiting ships. Some lived on the rewards for deserters, or found replacement crewmen either through persuasion or through shanghaiing. Many, such as David Whippy, also served as mediators between hostile native tribes as well as between natives and visiting ships. Whippy deserted his ship in 1820 and lived among the cannibal

Fijis for the rest of his life. The Fijis would sometimes capture the crew of a stranded ship for ransom, and eat them if they resisted. Whippy would try to rescue them but sometimes found only roasted bones. Ultimately he became British Consul to Fiji, and left many descendants among the islands.

There had always been a small number of castaways in the South Pacific since the earliest Spanish explorers, but the numbers increased dramatically in the early 19th century as the pre-requisite for a beachcombing in the Pacific was the era of whaling which began about 1819. It is estimated that 75% of beachcombers were sailors, particularly whalemen, who had jumped ship. They were predominantly British but with an increasing number of Americans, particularly in Hawaii and the Carolines. Perhaps 20% were English convicts who had been transported to Australia and escaped from the penal colonies there.

It is estimated that in 1850 there were over 2,000 beachcombers throughout Polynesia and Micronesia. The Polynesia and Melanesia communities were usually receptive to beachcombers and castaways who were absorbed into the local community, usually by formal adoption or by marriage, with the beachcombers and castaways often being considered a status symbol of the local chief. The social and commercial role of beachcombers ended when missionaries arrived and with the growth of a commercial community with European (palagi) traders, resident on each island, that were the representative of trading companies. Many beachcombers made the transition to become island traders.

Other Languages

In Uruguay, the term has been naturalized into the Spanish form *Bichicome*, and refers to poor or lower-class people. The Spanish form also draws on the similarities to the Spanish *bicho* (small animal/ insect) and *comer* (eat), Similarly, the term has entered the Greek slang through sailors, the word "pitsikómis" (ðéôóéêüìçò).

Archaeology

In archaeology the beachcombing lifestyle is associated with coastal shell-middens that sometimes accumulate over many hundreds if not thousands of years. Evidence at Klasies River Caves in South Africa, and Gulf of Zula in Eritrea, show that a beachcombing option is one of the earliest activities separating anatomically modern human *homo sapiens* from the ancestral subspecies of *Homo erectus*.

Modern Usage

Many modern beachcombers follow the "drift lines" or "tide lines" on the beach and are interested in the (mostly natural) objects that the sea casts up.

For these people, "beachcombing" is the recreational activity of looking for and finding various curiosities that have washed in with the tide: seashells of every kind, fossils, pottery shards, historical artifacts, sea beans (drift seeds), sea glass (beach glass) and driftwood.

Items such as lumber, plastics, and all manner of things that have been lost or discarded by seagoing vessels will be collected by some beachcombers, as long as the items are either decorative or useful in some way to the collector. (However, this usually does not include the great bulk of marine debris, most of which is neither useful nor decorative.)

Today, many people beachcomb for beach treasures and also because the activity offers them a natural prescription to achieve better emotional, physical and spiritual health.

Sophisticated recreational beachcombers use knowledge of how storms, geography, ocean currents, and seasonal events determine the arrival and exposure of rare finds. They also practice eco-conservation and do not kill mollusks for their shells, dig holes in the sand, or gouge cliff faces for fossils or reefs for coral specimens.

Many beachcombers serve as excellent stewards of the seashore, working with government agencies to monitor shore erosion, dumping and pollution, and reef and cliff damage, etc.

Recognized beachcomb experts today include oceanographer Curtis Ebbesmeyer (*Flotsametrics and the Floating World*); eco-educator Dr. Deacon Ritterbush (*A Beachcomber's Odyssey*); sea glass experts Richard LaMotte (*Pure Sea Glass*) and C.S. Lambert (*Sea Glass Chronicles*); geologist Margaret Carruthers (*Beach Stones*); shell specialists Chuck and Debbie Robinson (*The Art of Shelling*), and zoologist Dr. Blair Witherington and Dawn Witherington, (*Florida's Living Beaches: A Guide for the Curious Beachcomber*).

In Films and Television

Both the recreational and utilitarian aspects of beachcombing or "wrecking" were celebrated in the film *The Wrecking Season*, an award-winning film that portrays playwright Nick Darke's passion for beachcombing on the coast of Cornwall, UK.

A popular Canadian family television drama, *The Beachcombers*, focused on the work of beachcombers in late-twentieth-century British Columbia.

In the James Bond film 'On Her Majesty's Secret Service', Bond gets two weeks leave, and when Moneypenny asks him where he is off to, he replies: "just some place to laze about. Beachcombing".

Chenier

Chenier or Chénier is a sandy or shelly beach ridge that is part of a strand plain, called a "chenier plain," consisting of cheniers separated by intervening mud-flat deposits with marsh and swamp vegetation. Cheniers are typically 1 to 6 m high, tens of km long, hundreds of metres wide, and often wooded. Chenier plains can be tens of km wide. Cheniers and associated chenier plains are associated with shorelines characterized by generally low wave energy, low gradient, muddy shorelines, and abundant sediment supply. The name is derived from the French word for wood, "chêne," meaning oak, which grows on chenier ridges within southwest Louisiana.

Cheniers made primarily of shells of the Colorado Delta clam are found in the tidal flats of the Colorado River Delta in northeastern Baja California, Mexico, as well as in Essex, England. In Essex, Ground-truthing data demonstrate that the radar profiles accurately delineate the subsurface stratigraphy and sedimentary structure of the cheniers. Interpretation of the radar stratigraphy taken from the radar reflection profiles allows various deposits to be identified. These have resulted from overwashing, overtopping, sedimentation across the whole of a seaward dipping beachface or berm ridge welding onto the upper beachface. Each chenier is characterised by a different spatial arrangement of these four basic depositional units.

The Louisiana Chenier Plain, extending roughly from Sabine Lake to Vermilion Bay along the Gulf Coast, serves critical ecosystem functions, particularly as a wildlife habitat and stopover for migrating birds. As the region was settled, the cheniers were developed preferentially, owing to protection from flooding afforded by the higher elevation. Examples of communities situated on such ridges in the Chenier Plain include Grand Chenier in Cameron Parish, Pecan Island in Vermilion Parish and the former settlement of Cheniere Caminada adjacent to Grand Isle, south of New Orleans. The ridges, however, offer insufficient protection from storm surges, thus Cheniere Caminada was completely destroyed by a hurricane in 1893, and Pecan Island and Grand Chenier has been devastated by several powerful storms, including Hurricane Audrey, Hurricane Rita, and Hurricane Ike.

Kirby explains how sediment deposition through the null point hypothesis leads to the creation of chenier. Sediment fines are suspended and reworked aerially offshore leaving behind lag deposits of mainly bivalve and gastropod shells separated out from the finer substrate beneath, waves and currents then heap these deposits to form chenier ridges throughout the tidal zone which tend to be forced up the foreshore profile but also along the foreshore.

Logarithmic Spiral Beaches

A logarithmic spiral beach is a type of beach which develops in the direction under which it is sheltered by a headland, in an area called the shadow zone. It is characterized as a logarithmic spiral because if you look at it in plan view or aerially, it represents the same shape that is created from the logarithmic spiral function. These beaches are also commonly referred to as 'half heart' or 'crenulate' shaped bays, or 'headland bays'.

Logarithmic Spiral Function

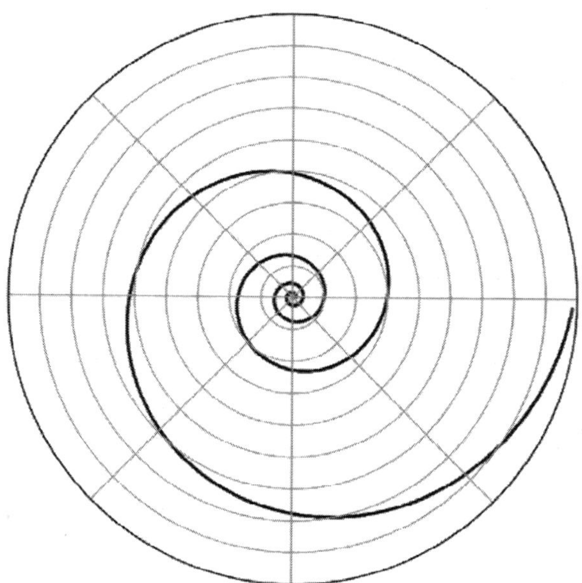

Figure: *Logarithmic Spiral*

The logarithmic spiral can be determined using the equation (written in polar coordinates):

$$r = e^{\theta \cot \alpha}$$

where:

θ = the angle of rotation, is located between two lines drawn from the origin to any two points on the spiral.

r = the ratio of the lengths between two lines that extend out from the origin. The two lines are given as R_0 and R. So r also equals the ratio R/R_0.

α = the angle between any line R from the origin and the line tangent to the spiral which is at the point where line R intersects the spiral. α is a constant for any given logarithmic spiral.

Spiral Development

This type of beach forms due to the refraction of approaching waves and their diffraction by an upcoast headland. The approaching wave front curves as a result of wave diffraction at the headland, which in turn causes the shoreline to bend and yield a log spiral shape. Log spiral beaches are often on swell-dominated coasts where waves generally approach the shoreline from one main direction at an oblique angle. The oblique approaching waves refract and diffract into the "shadow zone" which can be considered a relatively sheltered hook of beach behind the headland. Increase in sediment size, wave height, berm height, and swash zone gradient from the up coast headland generally characterizes the concave seaward curved part of the beach.

Beach Morphodynamics

Beach morphodynamics refers to the study of the interaction and adjustment of the seafloor topography and fluid hydrodynamic processes, seafloor morphologies and sequences of change dynamics involving the motion of sediment. Hydrodynamic processes include those of waves, tides and wind-induced currents.

While hydrodynamic processes respond instantaneously to morphological change, morphological change requires the redistribution of sediment. As sediment takes a finite time to move, there is a lag in the morphological response to hydrodynamic forcing. Sediment can therefore be considered to be a time-dependent coupling mechanism. Since the boundary conditions of hydrodynamic forcing change regularly, this may mean that the beach never attains equilibrium.

This systems approach to the coast was developed by Wright and Thom in 1977.

Examples of beach morphodynamic processes include the formation of beach cusps (on a very small scale), intertidal bars and log-spiral (Yasso 1965)/crenulate (Silvester 1980) embayments. Morphodynamic processes exhibit positive and negative feedbacks (such that beaches can, over different timescales, be considered to be both self-forcing and self-organised systems), nonlinearities and threshold behaviour.

According to their dynamic and morphological characteristics, exposed sandy beaches can be classified into several morphodynamic types (Short & Wright 1983, Short 1996). There is a large scale of morphodynamic states, this scale ranges from the dissipative state to the reflective extremes. Dissipative beaches are flat, have fine sand, incorporating waves that tend to break far from the intertidal zone and dissipate force progressively along wide surf zones. Reflective beaches are steep, and are known for their coarse sand; they have no surf zone, and the waves break brusquely on the intertidal zone.

Depending on beach state, near bottom currents show variations in the relative dominance of motions due to: incident waves, subharmonic oscillations, infragravity oscillations, and mean longshore and rip currents. On reflective beaches, incident waves and subharmonic edge waves are dominant. In highly dissipative surf zones, shoreward decay of incident waves is accompanied by shoreward growth of infragravity energy; in the inner surf zone, currents associated with infragravity standing waves dominate. On intermediate states with pronounced bar-trough (straight or crescentic) topographies, incident wave orbital velocities are generally dominant but significant roles are also played by subharmonic and infragravity standing waves, longshore currents, and rips. The strongest rips and associated feeder currents occur in association with intermediate transverse bar and rip topographies.(Virginia Institute of Marine Science- Scopus 460).

With respect to beach morphodynamics, a spectrum of morphodynamic states has been developed, ranging from reflective beaches at one end to dissipative beaches at the other. Reflective beaches are typically steep in profile with a narrow shoaling and surf zone, composed of coarse sediment, and characterised by surging breakers. Coarser sediment allows percolation during the swash part of the wave cycle, thus reducing the strength of backwash and allowing material to be deposited in the swash zone. At the other end of the continuum of beach states, dissipative beaches are wide and flat in profile, with a wide shoaling and surf zone, composed of finer sediment, and characterised by spilling breakers. Transitions between beach states are often caused by changes in wave energy, with storms causing reflective beach profiles to flatten (offshore movement of sediment under steeper waves), thus adopting a more dissipative profile.

Morphodynamic poos are also associated with other coastal landforms, for example spur and groove topography on coral reefs and tidal flats in infilling estuaries.

Boundary Current

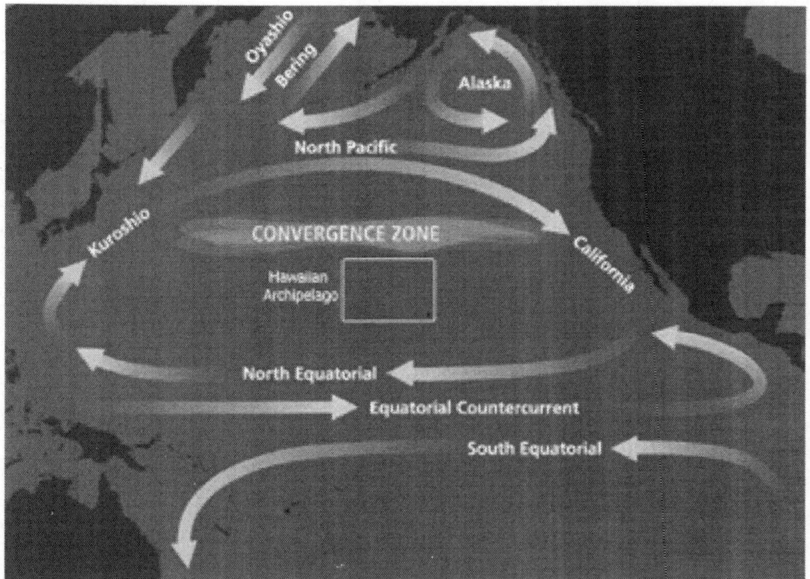

Figure: *The main ocean currents involved with the North Pacific Gyre*

Boundary currents are ocean currents with dynamics determined by the presence of a coastline, and fall into two distinct categories: western boundary currents and eastern boundary currents.

Eastern Boundary Currents

Eastern boundary currents are relatively shallow, broad and slow-flowing. They are found on the eastern side of oceanic basins (adjacent to the western coasts of continents). Subtropical eastern boundary currents flow equatorward, transporting cold water from higher latitudes to lower latitudes; examples include the Benguela Current, the Canary Current, the Peru Current, and the California Current. Coastal upwelling often brings nutrient-rich water into eastern boundary current regions, making them productive areas of the ocean.

Western Boundary Currents

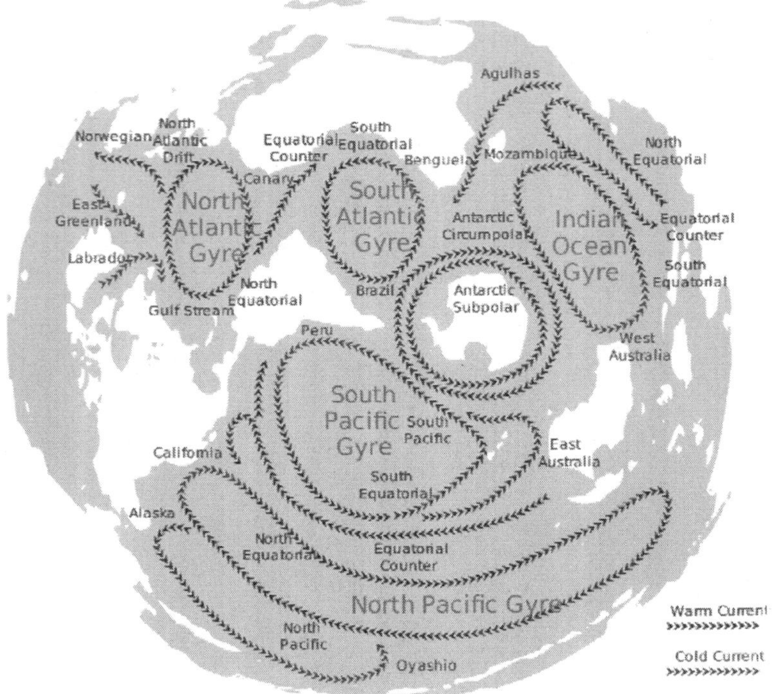

Figure: *The world's largest ocean gyres*

Western boundary currents are warm, deep, narrow, and fast flowing currents that form on the west side of ocean basins due to *western intensification*. They carry warm water from the tropics poleward. Examples include the Gulf Stream, the Agulhas Current, and the Kuroshio.

Western Intensification

Western intensification is the intensification of the western arm of an oceanic current, particularly a large gyre in an ocean basin. The trade winds blow westward in the tropics, and the westerlies blow eastward at mid-latitudes. This wind pattern applies a stress to the subtropical ocean surface with negative curl in the northern hemisphere and a positive curl in the southern hemisphere. The resulting Sverdrup transport is equatorward in both cases. Because of conservation of mass and potential vorticity conservation, that transport is balanced by a narrow, intense poleward current, which flows along the western boundary of the ocean basin, allowing the vorticity introduced by coastal friction to balance the vorticity input of the wind. Western intensification also occurs in the polar gyres, where the sign of the wind stress curl and the direction of the resulting currents are reversed. It is because of western intensification that the currents on the western boundary of a basin (such as the Gulf Stream, a current on the western side of the Atlantic Ocean) are stronger than those on the eastern boundary (such as the California Current, on the eastern side of the Pacific Ocean). Western intensification was first explained by the American oceanographer Henry Stommel.

In 1948, Henry Stommel published a paper in *Transactions, American Geophysical Union* titled "The Westward Intensification of Wind-Driven Ocean Currents", in which he used a simple, homogeneous, rectangular ocean model to examine the streamlines and surface height contours for an ocean at a non-rotating frame, an ocean characterized by a constant Coriolis parameter and finally, a real-case ocean basin with a latitudinally-varying Coriolis parameter. In this simple, modelling setting, the principal factors that were accounted for influencing the oceanic circulation were surface wind stress, bottom friction, a variable surface height leading to horizontal pressure gradients, and finally, the Coriolis effect.

In his simplified model, he assumed an ocean of constant density and constant depth D (when at rest), varying at increments of depth h in the presence of ocean currents; he also introduced a linearized, frictional term to account for the dissipative effects that prevent the real ocean from accelerating. He starts, thus, from the steady-state momentum and continuity equations:

$$f(D+h)v - Fcos(\frac{\pi y}{b}) - Ru - g(D+h)\frac{\partial h}{\partial x} = 0 \qquad (1)$$

$$-f(D+h)u - Rv - g(D+h)\frac{\partial h}{\partial y} = 0 \qquad (2)$$

$$\frac{\partial[(D+h)u]}{\partial x} + \frac{\partial[(D+h)v]}{\partial y} = 0 \qquad (3)$$

Multiplying (1) with $\frac{\partial}{\partial y}$ and (2) with $\frac{\partial}{\partial x}$ and using (3), he obtained the following equation (4), whose solutions for different ocean systems emphasize the role of the variation of the Coriolis parameter with latitude in inciting the strengthening of western boundary currents. Such currents are observed to be much faster, deeper, narrower and warmer than their eastern counterparts. Equation (4)

$$v(D+h)(\frac{\partial f}{\partial y})+\frac{\pi F}{b}sin(\frac{\pi y}{b})+R(\frac{\partial v}{\partial x}-\frac{\partial u}{\partial y})=0 \tag{4}$$

yields equation (5)

$$\nabla^2\psi+\alpha(\frac{\partial \psi}{\partial x})=\gamma sin(\frac{\pi y}{b}) \tag{5}$$

via the use of a Stream function ø, making the approximation D >> h and using the definitions

$$\alpha=(\frac{D}{R})(\frac{\partial f}{\partial y}) \quad \text{and} \quad \gamma=\frac{\pi F}{Rb}.$$

F is a constant used to represent wind stress via a simple functional form:

$$\frac{\pi F}{b}sin(\frac{\pi y}{b}).$$

One can obtain different solutions to the above equation, for different states of the modelled ocean; for a non-rotating state (zero Coriolis parameter) as well as an ocean state at which the Coriolis parameter is a constant, the ocean circulation does not demonstrate any preference towards intensification/acceleration near the western boundary.

The streamlines exhibit a symmetric behaviour in all directions, with the height contours demonstrating a nearly parallel relation to the streamlines, in the case of the homogeneously rotating ocean. Finally, for the case of interest - the one in which the Coriolis force is latitudinally variant - a distinct tendency for an asymmetrical streamline diagram is noted, with an observed, intense clustering towards the western part of the modelled ocean. A nice set of figures depicting the distribution of streamlines and height contours for the cases of a uniformly-rotating ocean and an ocean where the Coriolis force is linearly dependent on latitude.

Sverdrup Balance and Physics of Western Intensification

The physics of western intensification can be understood through a mechanism that helps maintain the vortex balance along an ocean gyre. Harald Sverdrup was the first one, preceding Henry Stommel, to attempt to explain the mid-ocean vorticity balance by looking at the relationship between surface wind forcings and the mass transport within the upper ocean layer. He assumed a geostrophic interior flow, while neglecting any frictional or viscosity effects and presuming that the circulation vanishes at some depth in the ocean. This prohibited the application of his theory to the western boundary currents, since some form of dissipative effect (bottom Ekman layer) would be later shown to be necessary to predict a closed circulation for an entire ocean basin and to counteract the wind-driven flow.

Sverdrup introduced a potential vorticity argument to connect the net, interior flow of the oceans to the surface wind stress and the incited planetary vorticity perturbations. For instance, Ekman convergence in the sub-tropics (related to the existence of the trade winds in the tropics and the westerlies in the mid-latitudes) was suggested to lead to a downward vertical velocity and therefore, a squashing of the water columns, which subsequently forces the ocean gyre to spin more slowly (via angular momentum conservation).

This is accomplished via a decrease in planetary vorticity (since relative vorticity variations are not significant in large ocean circulations), a phenomenon attainable through an equator-wardly directed, interior flow that characterizes the subtropical gyre. The opposite is applicable when Ekman divergence is induced, leading to Ekman absorption (suction) and a subsequent, water column stretching and poleward return flow, a characteristic of sub-polar gyres.

This return flow, as shown by Stommel, occurs in a meridional current, concentrated near the western boundary of an ocean basin. To balance the vorticity source induced by the wind stress forcing, Stommel introduced a linear frictional term in the Sverdrup equation, functioning as the vorticity sink.

This bottom ocean, frictional drag on the horizontal flow allowed Stommel to theoretically predict a closed, basin-wide circulation, while demonstrating the west-ward intensification of wind-driven gyres and its attribution to the Coriolis variation with latitude (beta effect). Walter Munk (1950) further implemented Stommel's theory of western intensification by using a more realistic frictional term, while emphasizing "the lateral dissipation of eddy energy."

In this way, not only did he reproduce Stommel's results, recreating thus the circulation of a western boundary current of an ocean gyre resembling the Gulf stream, but he also showed that sub-polar gyres should develop northward of the subtropical ones, spinning in the opposite direction.

Supralittoral Zone

The supralittoral zone, also known as the splash zone, spray zone or the supratidal zone, is the area above the spring high tide line, on coastlines and estuaries, that is regularly splashed, but not submerged by ocean water. Seawater penetrates these elevated areas only during storms with high tides.

Organisms here must cope also with exposure to air, fresh water from rain, cold, heat and predation by land animals and seabirds. At the top of this area, patches of dark lichens can appear as crusts on rocks. Some types of periwinkles, Neritidae and detritus feeding Isopoda commonly inhabit the lower supralitoral.

Mild-slope Equation

In fluid dynamics, the mild-slope equation describes the combined effects of diffraction and refraction for water waves propagating over bathymetry and due to lateral boundaries—like breakwaters and coastlines. It is an approximate model, deriving its name from being originally developed for wave propagation over mild slopes of the sea floor. The mild-slope equation is often used in coastal engineering to compute the wave-field changes near harbours and coasts.

The mild-slope equation models the propagation and transformation of water waves, as they travel through waters of varying depth and interact with lateral boundaries such as cliffs, beaches, seawalls and breakwaters. As a result, it describes the variations in wave amplitude, or equivalently wave height. From the wave amplitude, the amplitude of the flow velocity oscillations underneath the water surface can also be computed. These quantities—wave amplitude and flow-velocity amplitude—may subsequently be used to determine the wave effects on coastal and offshore structures, ships and other floating objects, sediment transport and resulting geomorphology changes of the sea bed and coastline, mean flow fields and mass transfer of dissolved and floating materials. Most often, the mild-slope equation is solved by computer using methods from numerical analysis.

A first form of the mild-slope equation was developed by Eckart in 1952, and an improved version—the mild-slope equation in its classical formulation—has been derived independently by Juri Berkhoff in 1972. Thereafter, many modified and extended forms have been proposed, to include the effects of, for instance: wave–current interaction, wave nonlinearity, steeper sea-bed slopes, bed friction and wave breaking. Also parabolic approximations to the mild-slope equation are often used, in order to reduce the computational cost.

In case of a constant depth, the mild-slope equation reduces to the Helmholtz equation for wave diffraction.

Formulation for Monochromatic Wave Motion

For monochromatic waves according to linear theory—with the free surface elevation given as $\zeta(x,y,t)=\Re\left\{\eta(x,y)e^{-i\omega t}\right\}$ and the waves propagating on a fluid layer of mean water depth $h(x,y)$—the mild-slope equation is:

$$\nabla\cdot\left(c_p c_g \nabla\eta\right)+k^2 c_p c_g \eta=0,$$

where:

- $\eta(x,y)$ is the complex-valued amplitude of the free-surface elevation $\zeta(x,y,t)$;
- (x,y) is the horizontal position;
- $\omega$is the angular frequency of the monochromatic wave motion;
- i is the imaginary unit;
- $\Re\{\cdot\}$ means taking the real part of the quantity between braces;
- ∇ is the horizontal gradient operator;
- $\nabla\cdot$ is the divergence operator;
- k is the wavenumber;
- c_p is the phase speed of the waves and
- c_g is the group speed of the waves.

The phase and group speed depend on the dispersion relation, and are derived from Airy wave theory as:

$$\omega^2 = gk\tanh(kh),$$

$$c_p = \frac{\omega}{k} \quad \text{and}$$

$$c_g = \frac{1}{2}c_p\left[1+kh\frac{1-\tanh^2(kh)}{\tanh(kh)}\right]$$

where

- g is Earth's gravity and
- $\tanh$ is the hyperbolic tangent.

For a given angular frequency ω, the wavenumber k has to be solved from the dispersion equation, which relates these two quantities to the water depth h.

Transformation to an Inhomogeneous Helmholtz Equation

Through the transformation

$$\psi=\eta\sqrt{c_p c_g},$$

the mild slope equation can be cast in the form of an inhomogeneous Helmholtz equation:

$$\Delta\psi+k_c^2\psi=0 \qquad \text{with} \qquad k_c^2=k^2-\frac{\Delta\left(\sqrt{c_p c_g}\right)}{\sqrt{c_p c_g}},$$

where Δ is the Laplace operator.

Applicability and Validity of the Mild-slope Equation

The standard mild slope equation, without extra terms for bed slope and bed curvature, provides accurate results for the wave field over bed slopes ranging from 0 to about 1/3. However, some subtle aspects, like the amplitude of reflected waves, can be completely wrong, even for slopes going to zero. This mathematical curiosity has little practical importance in general since this reflection becomes vanishingly small for small bottom slopes.

Low Marsh

Low marsh is a tidal marsh zone. It is characterized as being flooded daily. The water of the low marsh is shallow, silt laden, salty, and warm. These factors decrease the water's capacity to hold dissolved oxygen. Spartina alterniflora and some types of algae are the only plants that can survuve in this zone. Some areas of the low marsh are covered by the tides for most of the day. The mud that makes up the soil in the low marsh has high levels of salinity. A plant that lives in soil that is highly saline is known as a halophyte.

High Marsh

High marsh is a tidal marsh zone located above the Mean Highwater Mark (MHW) which, in contrast to the low marsh zone, is inundated infrequently during periods of extreme high tide and storm surge associated with coastal storms. The high marsh is the intermittent zone between the low marsh and the uplands, an entirely terrestrial area rarely flooded during events of extreme tidal action caused by severe coastal storms.

The high marsh is distinguished from the low marsh by its sandy soil and higher elevation. The elevation of the high marsh allows this zone to be covered by the high tide for no more than an hour a day. With the soil exposed to air for long periods of time, evaporation occurs, leading to high salinity levels, up to four times that of sea water. Areas of extremely high salinity prohibit plant growth all together. These barren sandy areas are known as "salt pans". Some cordgrass plants do survive here, but are stunted and do not reach their full size.

Tide

Tides are the rise and fall of sea levels caused by the combined effects of the gravitational forces exerted by the Moon and the Sun and the rotation of the Earth.

Some shorelines experience two almost equal high tides and two low tides each day, called a semi-diurnal tide. Some locations experience only one high and one low tide each day, called a diurnal tide. Some locations experience two uneven tides a day, or sometimes one high and one low each day; this is called a mixed tide.

The times and amplitude of the tides at a locale are influenced by the alignment of the Sun and Moon, by the pattern of tides in the deep ocean, by the amphidromic systems of the oceans, and by the shape of the coastline and near-shore bathymetry.

Tides vary on timescales ranging from hours to years due to numerous influences. To make accurate records, tide gauges at fixed stations measure the water level over time. Gauges ignore variations caused by waves with periods shorter than minutes.

These data are compared to the reference (or datum) level usually called mean sea level. While tides are usually the largest source of short-term sea-level fluctuations, sea levels are also subject to forces such as wind and barometric pressure changes, resulting in storm surges, especially in shallow seas and near coasts.

Tidal phenomena are not limited to the oceans, but can occur in other systems whenever a gravitational field that varies in time and space is present. For example, the solid part of the Earth is affected by tides, though this is not as easily seen as the water tidal movements.

Characteristics

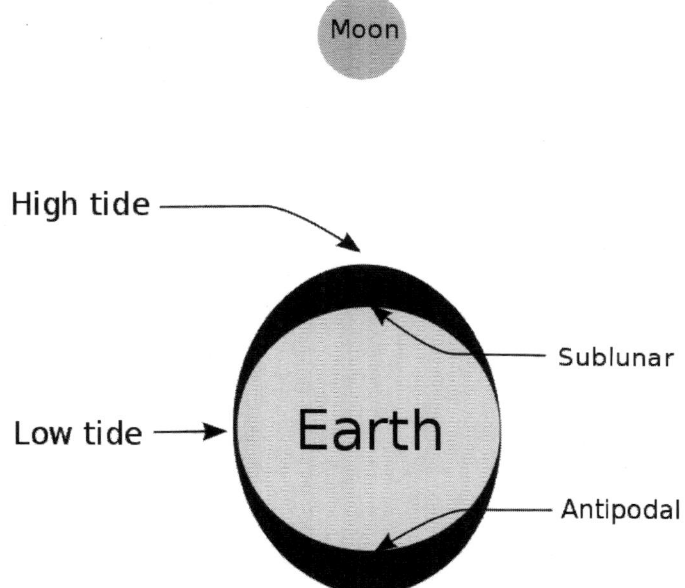

Tide changes proceed via the following stages:

- Sea level rises over several hours, covering the intertidal zone; flood tide.
- The water rises to its highest level, reaching high tide.
- Sea level falls over several hours, revealing the intertidal zone; ebb tide.
- The water stops falling, reaching low tide.

Tides produce oscillating currents known as tidal streams. The moment that the tidal current ceases is called slack water or slack tide. The tide then reverses direction and is said to be turning. Slack water usually occurs near high water and low water. But there are locations where the moments of slack tide differ significantly from those of high and low water.

Tides are most commonly *semi-diurnal* (two high waters and two low waters each day), or *diurnal* (one tidal cycle per day). The two high waters on a given day are typically not the same height (the daily inequality); these are the *higher high water* and the *lower high water* in tide tables. Similarly, the two low waters each day are the *higher low water* and the *lower low water.* The daily inequality is not consistent and is generally small when the Moon is over the equator.

Tidal Constituents

Tidal changes are the net result of multiple influences that act over varying periods. These influences are called tidal constituents. The primary constituents are the Earth's rotation, the positions of the Moon and the Sun relative to Earth, the Moon's altitude (elevation) above the Earth's equator, and bathymetry.

Variations with periods of less than half a day are called *harmonic constituents*. Conversely, cycles of days, months, or years are referred to as *long period* constituents.

The tidal forces affect the entire earth, but the movement of the solid Earth is only centimeters. The atmosphere is much more fluid and compressible so its surface moves kilometers, in the sense of the contour level of a particular low pressure in the outer atmosphere.

Principal Lunar Semi-diurnal Constituent

In most locations, the largest constituent is the "principal lunar semi-diurnal", also known as the *M2* (or M_2) tidal constituent. Its period is about 12 hours and 25.2 minutes, exactly half a *tidal lunar day*, which is the average time separating one lunar zenith from the next, and thus is the time required for the Earth to rotate once relative to the Moon. Simple tide clocks track this constituent. The lunar day is longer than the Earth day because the Moon orbits in the same direction the Earth spins. This is analogous to the minute hand on a watch crossing the hour hand at 12:00 and then again at about 1:05½ (not at 1:00).

The Moon orbits the Earth in the same direction as the Earth rotates on its axis, so it takes slightly more than a day—about 24 hours and 50 minutes—for the Moon to return to the same location in the sky. During this time, it has passed overhead (culmination) once and underfoot once (at an hour angle of 00:00 and 12:00 respectively), so in many places the period of strongest tidal forcing is the above mentioned, about 12 hours and 25 minutes. The moment of highest tide is not necessarily when the Moon is nearest to zenith or nadir, but the period of the forcing still determines the time between high tides.

Because the gravitational field created by the Moon weakens with distance from the Moon, it exerts a slightly stronger than average force on the side of the Earth facing the Moon, and a slightly weaker force on the opposite side. The Moon thus tends to "stretch" the Earth slightly along the line connecting the two bodies. The solid Earth deforms a bit, but ocean water, being fluid, is free to move much more in response to the tidal force, particularly horizontally. As the Earth rotates, the magnitude and direction of the tidal force at any particular point on the Earth's surface change constantly; although the ocean never reaches equilibrium—there is never time for the fluid to "catch up" to the state it would eventually reach if the tidal force were constant—the changing tidal force nonetheless causes rhythmic changes in sea surface height.

Semi-diurnal Range Differences

When there are two high tides each day with different heights (and two low tides also of different heights), the pattern is called a *mixed semi-diurnal tide.*

Range Variation: Springs and Neaps

The semi-diurnal range (the difference in height between high and low waters over about half a day) varies in a two-week cycle. Approximately twice a month, around new moon and full moon when the Sun, Moon, and Earth form a line (a condition known as syzygy), the tidal force due to the sun reinforces that due to the Moon. The tide's range is then at its maximum; this is called the *spring tide*, or just *springs*. It is not named after the season, but, like that word, derives from the meaning "jump, burst forth, rise", as in a natural spring.

When the Moon is at first quarter or third quarter, the sun and Moon are separated by 90° when viewed from the Earth, and the solar tidal force partially cancels the Moon's. At these points in the lunar cycle, the tide's range is at its minimum; this is called the *neap tide*, or *neaps* (a word of uncertain origin).

Spring tides result in high waters that are higher than average, low waters that are lower than average, 'slack water' time that is shorter than average, and stronger tidal currents than average. Neaps result in less-extreme tidal conditions. There is about a seven-day interval between springs and neaps.

Lunar Altitude

The changing distance separating the Moon and Earth also affects tide heights. When the Moon is closest, at perigee, the range increases, and when it is at apogee, the range shrinks. Every 7½ lunations (the full cycles from full moon to new to full), perigee coincides with either a new or full moon causing perigean spring tides with the largest *tidal range*. Even at its most powerful this force is still weak causing tidal differences of inches at most.

Bathymetry

The shape of the shoreline and the ocean floor changes the way that tides propagate, so there is no simple, general rule that predicts the time of high water from the Moon's position in the sky. Coastal characteristics such as underwater bathymetry and coastline shape mean that individual location characteristics affect tide forecasting; actual high water time and height may differ from model predictions due to the coastal morphology's effects on tidal flow. However, for a given location the relationship between lunar altitude and the time of high or low tide (the lunitidal interval) is relatively constant and predictable, as is the time of high or low tide relative to other points on the same coast. For example, the high tide at Norfolk, Virginia, predictably occurs approximately two and a half hours before the Moon passes directly overhead.

Land masses and ocean basins act as barriers against water moving freely around the globe, and their varied shapes and sizes affect the size of tidal frequencies. As a result, tidal patterns vary. For example, in the U.S., the East coast has predominantly semi-diurnal tides, as do Europe's Atlantic coasts, while the West coast predominantly has mixed tides.

Other Constituents

These include solar gravitational effects, the obliquity (tilt) of the Earth's equator and rotational axis, the inclination of the plane of the lunar orbit and the elliptical shape of the Earth's orbit of the sun.

A compound tide (or overtide) results from the shallow-water interaction of its two parent waves.

Phase and Amplitude

Because the M_2 tidal constituent dominates in most locations, the stage or *phase* of a tide, denoted by the time in hours after high water, is a useful concept. Tidal stage is also measured in degrees, with 360° per tidal cycle. Lines of constant tidal phase are called *cotidal lines*, which are analogous to contour lines of constant altitude on topographical maps. High water is reached simultaneously along the cotidal lines extending from the coast out into the ocean, and cotidal lines (and hence tidal phases) advance along the coast. Semi-diurnal and long phase constituents are measured from high water, diurnal from maximum flood tide. This and the discussion that follows is precisely true only for a single tidal constituent.

For an ocean in the shape of a circular basin enclosed by a coastline, the *cotidal lines* point radially inward and must eventually meet at a common point, the amphidromic point. The amphidromic point is at once cotidal with high and low waters, which is satisfied by *zero* tidal motion. (The rare exception occurs when the tide encircles an island, as it does around New Zealand, Iceland and Madagascar.) Tidal motion generally lessens moving away from continental coasts, so that crossing the cotidal lines are contours of constant *amplitude* (half the distance between high and low water) which decrease to zero at the amphidromic point.

For a semi-diurnal tide the amphidromic point can be thought of roughly like the centre of a clock face, with the hour hand pointing in the direction of the high water cotidal line, which is directly opposite the low water cotidal line. High water rotates about the amphidromic point once every 12 hours in the direction of rising cotidal lines, and away from ebbing cotidal lines.

This rotation is generally clockwise in the southern hemisphere and counterclockwise in the northern hemisphere, and is caused by the Coriolis effect. The difference of cotidal phase from the phase of a reference tide is the *epoch*. The reference tide is the hypothetical constituent "equilibrium tide" on a landless Earth measured at 0° longitude, the Greenwich meridian.

In the North Atlantic, because the cotidal lines circulate counterclockwise around the amphidromic point, the high tide passes New York Harbor approximately an hour ahead of Norfolk Harbor. South of Cape Hatteras the tidal forces are more complex, and cannot be predicted reliably based on the North Atlantic cotidal lines.

Physics

History of Tidal Physics: Investigation into tidal physics was important in the early development of heliocentrism and celestial mechanics, with the existence of two daily tides being explained by the Moon's gravity. Later the daily tides were explained more precisely by the interaction of the Moon's and the sun's gravity.

Galileo Galilei in his 1632 *Dialogue Concerning the Two Chief World Systems*, whose working title was *Dialogue on the Tides*, gave an explanation of the tides. The resulting theory, however, was incorrect as he attributed the tides to the sloshing of water caused by the Earth's movement around the sun. He hoped to provide mechanical proof of the Earth's movement – the value of his tidal theory is disputed. At the same time Johannes Kepler correctly suggested that the Moon caused the tides, which he based upon ancient observations and correlations, an explanation which was rejected by Galileo. It was originally mentioned in Ptolemy's Tetrabiblos as having derived from ancient observation.

Isaac Newton (1642–1727) was the first person to explain tides as the product of the gravitational attraction of astronomical masses. His explanation of the tides (and many other phenomena) was published in the *Principia* (1687) and used his theory of universal gravitation to explain the lunar and solar attractions as the origin of the tide-generating forces. Newton and others before Pierre-Simon Laplace worked the problem from the perspective of a static system (equilibrium theory), that provided an approximation that described the tides that would occur in a non-inertial ocean evenly covering the whole Earth.

The tide-generating force (or its corresponding potential) is still relevant to tidal theory, but as an intermediate quantity (forcing function) rather than as a final result; theory must also consider the Earth's accumulated dynamic tidal response to the applied forces, which response is influenced by bathymetry, Earth's rotation, and other factors.

In 1740, the Académie Royale des Sciences in Paris offered a prize for the best theoretical essay on tides. Daniel Bernoulli, Leonhard Euler, Colin Maclaurin and Antoine Cavalleri shared the prize.

Maclaurin used Newton's theory to show that a smooth sphere covered by a sufficiently deep ocean under the tidal force of a single deforming body is a prolate spheroid (essentially a three dimensional oval) with major axis directed towards the deforming body. Maclaurin was the first to write about the Earth's rotational effects on motion. Euler realized that the tidal force's *horizontal* component (more than the vertical) drives the tide. In 1744 Jean le Rond d'Alembert studied tidal equations for the atmosphere which did not include rotation.

Pierre-Simon Laplace formulated a system of partial differential equations relating the ocean's horizontal flow to its surface height, the first major dynamic theory for water tides. The Laplace tidal equations are still in use today. William Thomson, 1st Baron Kelvin, rewrote Laplace's equations in terms of vorticity which allowed for solutions describing tidally driven coastally trapped waves, known as Kelvin waves.

Others including Kelvin and Henri Poincaré further developed Laplace's theory. Based on these developments and the lunar theory of E W Brown describing the motions of the Moon, Arthur Thomas Doodson developed and published in 1921 the first modern development of the tide-generating potential in harmonic form: Doodson distinguished 388 tidal frequencies. Some of his methods remain in use.

Forces

The tidal force produced by a massive object (Moon, hereafter) on a small particle located on or in an extensive body (Earth, hereafter) is the vector difference between the gravitational force exerted by the Moon on the particle, and the gravitational force that would be exerted on the particle if it were located at the Earth's centre of mass. The solar *gravitational force* on the Earth is on average 179 times stronger than the lunar, but because the Sun is on average 389 times farther from the Earth, its field gradient is weaker. The solar tidal force is 46% as large as the lunar. More precisely, the lunar tidal acceleration (along the Moon–Earth axis, at the Earth's surface) is about 1.1×10^{-7} g, while the solar tidal acceleration (along the Sun–Earth axis, at the Earth's surface) is about 0.52×10^{-7} g, where g is the gravitational acceleration at the Earth's surface. Venus has the largest effect of the other planets, at 0.000113 times the solar effect.

The ocean's surface is closely approximated by an equipotential surface, (ignoring ocean currents) commonly referred to as the geoid. Since the gravitational force is equal to the potential's gradient, there are no tangential forces on such a surface, and the ocean surface is thus in gravitational equilibrium. Now consider the effect of massive external bodies such as the Moon and Sun. These bodies have strong gravitational fields that diminish with distance in space and which act to alter the shape of an equipotential surface on the Earth. This deformation has a fixed spatial orientation relative to the influencing body. The Earth's rotation relative to this shape causes the daily tidal cycle. Gravitational forces follow an inverse-square law (force is inversely proportional to the square of the distance), but tidal forces are inversely proportional to the cube of the distance. The ocean surface moves because of the changing tidal equipotential, rising when the tidal potential is high, which occurs on the parts of the Earth nearest to and furthest from the Moon. When the tidal equipotential changes, the ocean surface is no longer aligned with it, so the apparent direction of the vertical shifts. The surface then experiences a down slope, in the direction that the equipotential has risen.

Laplace's Tidal Equations

Ocean depths are much smaller than their horizontal extent. Thus, the response to tidal forcing can be modelled using the Laplace tidal equations which incorporate the following features:

1. The vertical (or radial) velocity is negligible, and there is no vertical shear—this is a sheet flow.
2. The forcing is only horizontal (tangential).
3. The Coriolis effect appears as an inertial force (fictitious) acting laterally to the direction of flow and proportional to velocity.
4. The surface height's rate of change is proportional to the negative divergence of velocity multiplied by the depth. As the horizontal velocity stretches or compresses the ocean as a sheet, the volume thins or thickens, respectively.

The boundary conditions dictate no flow across the coastline and free slip at the bottom.

The Coriolis effect (inertial force) steers currents moving towards the equator to the west and towards the east for flows moving away from the equator, allowing coastally trapped waves. Finally, a dissipation term can be added which is an analog to viscosity.

Amplitude and Cycle Time

The theoretical amplitude of oceanic tides caused by the moon is about 54 centimetres (21 in) at the highest point, which corresponds to the amplitude that would be reached if the ocean possessed a uniform depth, there were no landmasses, and the Earth were rotating in step with the moon's orbit. The sun similarly causes tides, of which the theoretical amplitude is about 25 centimetres (9.8 in) (46% of that of the moon) with a cycle time of 12 hours. At spring tide the two effects add to each other to a theoretical level of 79 centimetres (31 in), while at neap tide the theoretical level is reduced to 29 centimetres (11 in). Since the orbits of the Earth about the sun, and the moon about the Earth, are elliptical, tidal amplitudes change somewhat as a result of the varying Earth–sun and Earth–moon distances. This causes a variation in the tidal force and theoretical amplitude of about ±18% for the moon and ±5% for the sun. If both the sun and moon were at their closest positions and aligned at new moon, the theoretical amplitude would reach 93 centimetres (37 in).

Real amplitudes differ considerably, not only because of depth variations and continental obstacles, but also because wave propagation across the ocean has a natural period of the same order of magnitude as the rotation period: if there were no land masses, it would take about 30 hours for a long wavelength surface wave to propagate along the equator halfway around the Earth (by comparison, the Earth's lithosphere has a natural period of about 57 minutes). Earth tides, which raise and lower the bottom of the ocean, and the tide's own gravitational self attraction are both significant and further complicate the ocean's response to tidal forces.

Dissipation

Earth's tidal oscillations introduce dissipation at an average rate of about 3.75 terawatt. About 98% of this dissipation is by marine tidal movement. Dissipation arises as basin-scale tidal flows drive smaller-scale flows which experience turbulent dissipation. This tidal drag creates torque on the moon that gradually transfers angular momentum to its orbit, and a gradual increase in Earth–moon separation. The equal and opposite torque on the Earth correspondingly decreases its rotational velocity. Thus, over geologic time, the moon recedes from the Earth, at about 3.8 centimetres (1.5 in)/year, lengthening the terrestrial day. Day length has increased by about 2 hours in the last 600 million years. Assuming (as a crude approximation) that the deceleration rate has been constant, this would imply that 70 million years ago, day length was on the order of 1% shorter with about 4 more days per year.

Observation and Prediction

History

From ancient times, tidal observation and discussion has increased in sophistication, first marking the daily recurrence, then tides' relationship to the sun and moon. Pytheas travelled to the British Isles about 325 BC and seems to be the first to have related spring tides to the phase of the moon. In the 2nd century BC, the Babylonian astronomer, Seleucus of Seleucia, correctly described the phenomenon of tides in order to support his heliocentric theory. He correctly theorized that tides were caused by the moon, although he believed that the interaction was mediated by the pneuma. He noted that tides varied in time and strength in different parts of the world. According to Strabo (1.1.9), Seleucus was the first to link tides to the lunar attraction, and that the height of the tides depends on the moon's position relative to the sun.

The *Naturalis Historia* of Pliny the Elder collates many tidal observations, e.g., the spring tides are a few days after (or before) new and full moon and are highest around the equinoxes, though Pliny noted many relationships now regarded as fanciful. In his *Geography*, Strabo described tides in the Persian Gulf having their greatest range when the moon was furthest from the plane of the equator. All this despite the relatively small amplitude of Mediterranean basin tides. (The strong currents through the Euripus Strait and the Strait of Messina puzzled Aristotle.) Philostratus discussed tides in Book Five of *The Life of Apollonius of Tyana*. Philostratus mentions the moon, but attributes tides to "spirits". In Europe around 730 AD, the Venerable Bede described how the rising tide on one coast of the British Isles coincided with the fall on the other and described the time progression of high water along the Northumbrian coast.

The first tide table in China was recorded in 1056 AD primarily for visitors wishing to see the famous tidal bore in the Qiantang River. The first known British tide table is thought to be that of John Wallingford, who died Abbot of St. Albans in 1213, based on high water occurring 48 minutes later each day, and three hours earlier at the Thames mouth than upriver at London.

William Thomson (Lord Kelvin) led the first systematic harmonic analysis of tidal records starting in 1867. The main result was the building of a tide-predicting machine using a system of pulleys to add together six harmonic time functions. It was "programmed" by resetting gears and chains to adjust phasing and amplitudes. Similar machines were used until the 1960s.

The first known sea-level record of an entire spring–neap cycle was made in 1831 on the Navy Dock in the Thames Estuary. Many large ports had automatic tide gage stations by 1850.

William Whewell first mapped co-tidal lines ending with a nearly global chart in 1836. In order to make these maps consistent, he hypothesized the existence of amphidromes where co-tidal lines meet in the mid-ocean. These points of no tide were confirmed by measurement in 1840 by Captain Hewett, RN, from careful soundings in the North Sea.

Timing

The tidal forces due to the Moon and Sun generate very long waves which travel all around the ocean following the paths shown in co-tidal charts. The time when the crest of the wave reaches a port then gives the time of high water at the port. The time taken for the wave to travel around the ocean also means that there is a delay between the phases the moon and their effect on the tide. Springs and neaps in the North Sea, for example, are two days behind the new/full moon and first/third quarter moon. This is called the tide's *age*.

The ocean bathymetry greatly influences the tide's exact time and height at a particular coastal point. There are some extreme cases; the Bay of Fundy, on the east coast of Canada, is often stated to have the world's highest tides because of its shape, bathymetry, and its distance from the continental shelf edge. Measurements made in November 1998 at Burntcoat Head in the Bay of Fundy recorded a maximum range of 16.3 metres (53 ft) and a highest predicted extreme of 17 metres (56 ft). Similar measurements made in March 2002 at Leaf Basin, Ungava Bay in northern Quebec gave similar values (allowing for measurement errors), a maximum range of 16.2 metres (53 ft) and a highest predicted extreme of 16.8 metres (55 ft). Ungava Bay and the Bay of Fundy lie similar distances from the continental shelf edge, but Ungava Bay is free of pack ice for only about four months every year while the Bay of Fundy rarely freezes.

Southampton in the United Kingdom has a double high water caused by the interaction between the region's different tidal harmonics, caused primarily by the east/west orientation of the English Channel and the fact that when it is high water at Dover it is low water at Land's End (some 300 nautical miles distant) and vice versa. This is contrary to the popular belief that the flow of water around the Isle of Wight creates two high waters. The Isle of Wight is important, however, since it is responsible for the 'Young Flood Stand', which describes the pause of the incoming tide about three hours after low water.

Because the oscillation modes of the Mediterranean Sea and the Baltic Sea do not coincide with any significant astronomical forcing period, the largest tides are close to their narrow connections with the Atlantic Ocean. Extremely small tides also occur for the same reason in the Gulf of Mexico and Sea of Japan. Elsewhere, as along the southern coast of Australia, low tides can be due to the presence of a nearby amphidrome.

Analysis

Isaac Newton's theory of gravitation first enabled an explanation of why there were generally two tides a day, not one, and offered hope for detailed understanding. Although it may seem that tides could be predicted via a sufficiently detailed knowledge of the instantaneous astronomical forcings, the actual tide at a given location is determined by astronomical forces accumulated over many days. Precise results require detailed knowledge of the shape of all the ocean basins—their bathymetry and coastline shape.

Current procedure for analysing tides follows the method of harmonic analysis introduced in the 1860s by William Thomson. It is based on the principle that the astronomical theories of the motions of sun and moon determine a large number of component frequencies, and at each frequency there is a component of force tending to produce tidal motion, but that at each place of interest on the Earth, the tides respond at each frequency with an amplitude and phase peculiar to that locality. At each place of interest, the tide heights are therefore measured for a period of time sufficiently long (usually more than a year in the case of a new port not previously studied) to enable the response at each significant tide-generating frequency to be distinguished by analysis, and to extract the tidal constants for a sufficient number of the strongest known components of the astronomical tidal forces to enable practical tide prediction. The tide heights are expected to follow the tidal force, with a constant amplitude and phase delay for each component. Because astronomical frequencies and phases can be calculated with certainty, the tide height at other times can then be predicted once the response to the harmonic components of the astronomical tide-generating forces has been found.

The main patterns in the tides are

- the twice-daily variation
- the difference between the first and second tide of a day
- the spring–neap cycle
- the annual variation

The *Highest Astronomical Tide* is the perigean spring tide when both the sun and the moon are closest to the Earth. When confronted by a periodically varying function, the standard approach is to employ

Fourier series, a form of analysis that uses sinusoidal functions as a *basis* set, having frequencies that are zero, one, two, three, etc. times the frequency of a particular fundamental cycle. These multiples are called *harmonics* of the fundamental frequency, and the process is termed harmonic analysis. If the basis set of sinusoidal functions suit the behaviour being modelled, relatively few harmonic terms need to be added. Orbital paths are very nearly circular, so sinusoidal variations are suitable for tides.

For the analysis of tide heights, the Fourier series approach has in practice to be made more elaborate than the use of a single frequency and its harmonics. The tidal patterns are decomposed into many sinusoids having many fundamental frequencies, corresponding (as in the lunar theory) to many different combinations of the motions of the Earth, the moon, and the angles that define the shape and location of their orbits.

For tides, then, *harmonic analysis* is not limited to harmonics of a single frequency. In other words, the harmonies are multiples of many fundamental frequencies, not just of the fundamental frequency of the simpler Fourier series approach. Their representation as a Fourier series having only one fundamental frequency and its (integer) multiples would require many terms, and would be severely limited in the time-range for which it would be valid.

The study of tide height by harmonic analysis was begun by Laplace, William Thomson (Lord Kelvin), and George Darwin. A.T. Doodson extended their work, introducing the *Doodson Number* notation to organise the hundreds of resulting terms. This approach has been the international standard ever since, and the complications arise as follows: the tide-raising force is notionally given by sums of several terms. Each term is of the form

$$A \cdot \cos(w \cdot t + p)$$

where A is the amplitude, w is the angular frequency usually given in degrees per hour corresponding to t measured in hours, and p is the phase offset with regard to the astronomical state at time $t = 0$. There is one term for the moon and a second term for the sun. The phase p of the first harmonic for the moon term is called the lunitidal interval or high water interval. The next step is to accommodate the harmonic terms due to the elliptical shape of the orbits. Accordingly, the value of A is not a constant but also varying with time, slightly, about some average figure. Replace it then by $A(t)$ where A is another sinusoid, similar to the cycles and epicycles of Ptolemaic theory. Accordingly,

$$A(t) = A \cdot (1 + A_a \cdot \cos(w_a \cdot t + p_a)) \,,$$

which is to say an average value A with a sinusoidal variation about it of magnitude A_a, with frequency w_a and phase p_a. Thus the simple term is now the product of two cosine factors:

$$A \cdot [1 + A_a \cdot \cos(w_a \cdot t + p_a)] \cdot \cos(w \cdot t + p)$$

Given that for any x and y

$$\cos(x) \cdot \cos(y) = ½ \cdot \cos(x + y) + ½ \cdot \cos(x - y) \,,$$

it is clear that a compound term involving the product of two cosine terms each with their own frequency is the same as *three* simple cosine terms that are to be added at the original frequency and also at frequencies which are the sum and difference of the two frequencies of the product term. (Three, not two terms, since the whole expression is $(1 + \cos(x)) \cdot \cos(y)$.) Consider further that the tidal force on a location depends also on whether the moon (or the sun) is above or below the plane of the equator, and that these attributes have their own periods also incommensurable with a day and a month, and it is clear that many combinations result. With a careful choice of the basic astronomical frequencies, the Doodson Number annotates the particular additions and differences to form the frequency of each simple cosine term.

Remember that astronomical tides do *not* include weather effects. Also, changes to local conditions (sandbank movement, dredging harbour mouths, etc.) away from those prevailing at the measurement time affect the tide's actual timing and magnitude. Organisations quoting a "highest astronomical tide" for some location may exaggerate the figure as a safety factor against analytical uncertainties, distance from the nearest measurement point, changes since the last observation time, ground subsidence, etc., to avert liability should an engineering work be overtopped. Special care is needed when assessing the size of a "weather surge" by subtracting the astronomical tide from the observed tide.

Careful Fourier data analysis over a nineteen-year period (the *National Tidal Datum Epoch* in the U.S.) uses frequencies called the *tidal harmonic constituents.* Nineteen years is preferred because the Earth, moon and sun's relative positions repeat almost exactly in the Metonic cycle of 19 years, which is long enough to include the 18.613 year lunar nodal tidal constituent. This analysis can be done using only the knowledge of the forcing *period*, but without detailed understanding of the mathematical derivation, which means that useful tidal tables have been constructed for centuries.

The resulting amplitudes and phases can then be used to predict the expected tides. These are usually dominated by the constituents near 12 hours (the *semi-diurnal* constituents), but there are major constituents near 24 hours (*diurnal*) as well. Longer term constituents are 14 day or *fortnightly*, monthly, and semiannual. Semi-diurnal tides dominated coastline, but some areas such as the South China Sea and the Gulf of Mexico are primarily diurnal. In the semi-diurnal areas, the primary constituents M_2 (lunar) and S_2 (solar) periods differ slightly, so that the relative phases, and thus the amplitude of the combined tide, change fortnightly (14 day period).

In the M_2 plot above, each cotidal line differs by one hour from its neighbours, and the thicker lines show tides in phase with equilibrium at Greenwich. The lines rotate around the amphidromic points counterclockwise in the northern hemisphere so that from Baja California Peninsula to Alaska and from France to Ireland the M_2 tide propagates northward. In the southern hemisphere this direction is clockwise. On the other hand M_2 tide propagates counterclockwise around New Zealand, but this is because the islands act as a dam and permit the tides to have different heights on the islands' opposite sides. (The tides do propagate northward on the east side and southward on the west coast, as predicted by theory.)

The exception is at Cook Strait where the tidal currents periodically link high to low water. This is because cotidal lines 180° around the amphidromes are in opposite phase, for example high water across from low water at each end of Cook Strait. Each tidal constituent has a different pattern of amplitudes, phases, and amphidromic points, so the M_2 patterns cannot be used for other tide components.

Example Calculation

Because the moon is moving in its orbit around the earth and in the same sense as the Earth's rotation, a point on the earth must rotate slightly further to catch up so that the time between semidiurnal tides is not twelve but 12.4206 hours—a bit over twenty-five minutes extra. The two peaks are not equal. The two high tides a day alternate in maximum heights: lower high (just under three feet), higher high (just over three feet), and again lower high. Likewise for the low tides. When the Earth, moon, and sun are in line (sun–Earth–moon, or sun–

moon–Earth) the two main influences combine to produce spring tides; when the two forces are opposing each other as when the angle moon–Earth–sun is close to ninety degrees, neap tides result. As the moon moves around its orbit it changes from north of the equator to south of the equator. The alternation in high tide heights becomes smaller, until they are the same (at the lunar equinox, the moon is above the equator), then redevelop but with the other polarity, waxing to a maximum difference and then waning again.

Current

The tides' influence on current flow is much more difficult to analyse, and data is much more difficult to collect. A tidal height is a simple number which applies to a wide region simultaneously. A flow has both a magnitude and a direction, both of which can vary substantially with depth and over short distances due to local bathymetry. Also, although a water channel's centre is the most useful measuring site, mariners object when current-measuring equipment obstructs waterways. A flow proceeding up a curved channel is the same flow, even though its direction varies continuously along the channel. Surprisingly, flood and ebb flows are often not in opposite directions. Flow direction is determined by the upstream channel's shape, not the downstream channel's shape. Likewise, eddies may form in only one flow direction.

Nevertheless, current analysis is similar to tidal analysis: in the simple case, at a given location the flood flow is in mostly one direction, and the ebb flow in another direction. Flood velocities are given positive sign, and ebb velocities negative sign. Analysis proceeds as though these are tide heights.

In more complex situations, the main ebb and flood flows do not dominate. Instead, the flow direction and magnitude trace an ellipse over a tidal cycle (on a polar plot) instead of along the ebb and flood lines. In this case, analysis might proceed along pairs of directions, with the primary and secondary directions at right angles. An alternative is to treat the tidal flows as complex numbers, as each value has both a magnitude and a direction.

Tide flow information is most commonly seen on nautical charts, presented as a table of flow speeds and bearings at hourly intervals, with separate tables for spring and neap tides. The timing is relative to high water at some harbour where the tidal behaviour is similar in pattern, though it may be far away.

As with tide height predictions, tide flow predictions based only on astronomical factors do not incorporate weather conditions, which can *completely* change the outcome.

The tidal flow through Cook Strait between the two main islands of New Zealand is particularly interesting, as the tides on each side of the strait are almost exactly out of phase, so that one side's high water is simultaneous with the other's low water. Strong currents result, with almost zero tidal height change in the strait's centre. Yet, although the tidal surge normally flows in one direction for six hours and in the reverse direction for six hours, a particular surge might last eight or ten hours with the reverse surge enfeebled. In especially boisterous weather conditions, the reverse surge might be entirely overcome so that the flow continues in the same direction through three or more surge periods.

A further complication for Cook Strait's flow pattern is that the tide at the north side (e.g. at Nelson) follows the common bi-weekly spring–neap tide cycle (as found along the west side of the country), but the south side's tidal pattern has only *one* cycle per month, as on the east side: Wellington, and Napier.

The graph of Cook Strait's tides shows separately the high water and low water height and time, through November 2007; these are *not* measured values but instead are calculated from tidal parameters derived from years-old measurements. Cook Strait's nautical chart offers tidal current information. For instance the January 1979 edition for 41°13 ·9'S 174°29 ·6'E (north west of Cape Terawhiti) refers timings to Westport while the January 2004 issue refers to Wellington. Near Cape Terawhiti in the middle of Cook Strait the tidal height variation is almost nil while the tidal current reaches its maximum, especially near the notorious Karori Rip. Aside from weather effects, the actual currents through Cook Strait are influenced by the tidal height differences between the two ends of the strait and as can be seen, only one of the two spring tides at the north end (Nelson) has a counterpart spring tide at the south end (Wellington), so the resulting behaviour follows neither reference harbour.

Power Generation

Tidal energy can be extracted by two means: inserting a water turbine into a tidal current, or building ponds that release/admit water through a turbine. In the first case, the energy amount is entirely determined by the timing and tidal current magnitude.

However, the best currents may be unavailable because the turbines would obstruct ships. In the second, the impoundment dams are expensive to construct, natural water cycles are completely disrupted, ship navigation is disrupted. However, with multiple ponds, power can be generated at chosen times. So far, there are few installed systems for tidal power generation (most famously, La Rance by Saint Malo, France) which faces many difficulties. Aside from environmental issues, simply withstanding corrosion and biological fouling pose engineering challenges.

Tidal power proponents point out that, unlike wind power systems, generation levels can be reliably predicted, save for weather effects. While some generation is possible for most of the tidal cycle, in practice turbines lose efficiency at lower operating rates. Since the power available from a flow is proportional to the cube of the flow speed, the times during which high power generation is possible are brief.

Navigation

Tidal flows are important for navigation, and significant errors in position occur if they are not accommodated. Tidal heights are also important; for example many rivers and harbours have a shallow "bar" at the entrance which prevents boats with significant draft from entering at low tide. Until the advent of automated navigation, competence in calculating tidal effects was important to naval officers. The certificate of examination for lieutenants in the Royal Navy once declared that the prospective officer was able to "shift his tides".

Tidal flow timings and velocities appear in *tide charts* or a tidal stream atlas. Tide charts come in sets. Each chart covers a single hour between one high water and another (they ignore the leftover 24 minutes) and show the average tidal flow for that hour. An arrow on the tidal chart indicates the direction and the average flow speed (usually in knots) for spring and neap tides. If a tide chart is not available, most nautical charts have "tidal diamonds" which relate specific points on the chart to a table giving tidal flow direction and speed.

The standard procedure to counteract tidal effects on navigation is to

(1) calculate a "dead reckoning" position (or DR) from travel distance and direction,

(2) mark the chart (with a vertical cross like a plus sign) and

(3) draw a line from the DR in the tide's direction.

The distance the tide moves the boat along this line is computed by the tidal speed, and this gives an "estimated position" or EP (traditionally marked with a dot in a triangle).

Nautical charts display the water's "charted depth" at specific locations with "soundings" and the use of bathymetric contour lines to depict the submerged surface's shape. These depths are relative to a "chart datum", which is typically the water level at the lowest possible astronomical tide (although other datums are commonly used, especially historically, and tides may be lower or higher for meteorological reasons) and are therefore the minimum possible water depth during the tidal cycle. "Drying heights" may also be shown on the chart, which are the heights of the exposed seabed at the lowest astronomical tide.

Tide tables list each day's high and low water heights and times. To calculate the actual water depth, add the charted depth to the published tide height. Depth for other times can be derived from tidal curves published for major ports. The rule of twelfths can suffice if an accurate curve is not available. This approximation presumes that the increase in depth in the six hours between low and high water is: first hour — 1/12, second — 2/12, third — 3/12, fourth — 3/12, fifth — 2/12, sixth — 1/12.

Biological Aspects

Intertidal ecology: Intertidal ecology is the study of intertidal ecosystems, where organisms live between the low and high water lines. At low water, the intertidal is exposed (or 'emersed') whereas at high water, the intertidal is underwater (or 'immersed'). Intertidal ecologists therefore study the interactions between intertidal organisms and their environment, as well as among the different species. The most important interactions may vary according to the type of intertidal community. The broadest classifications are based on substrates — rocky shore or soft bottom.

Intertidal organisms experience a highly variable and often hostile environment, and have adapted to cope with and even exploit these conditions. One easily visible feature is vertical zonation, in which the community divides into distinct horizontal bands of specific species at each elevation above low water. A species' ability to cope with desiccation determines its upper limit, while competition with other species sets its lower limit. Humans use intertidal regions for food and recreation. Overexploitation can damage intertidals directly. Other anthropogenic actions such as introducing invasive species and climate change have large negative effects. Marine Protected Areas are one option communities can apply to protect these areas and aid scientific research.

Biological Rhythms

The approximately fortnightly tidal cycle has large effects on intertidal and marine organisms. Hence their biological rhythms tend to occur in rough multiples of this period. Many other animals such as the vertebrates, display similar rhythms. Examples include gestation and egg hatching. In humans, the menstrual cycle lasts roughly a lunar month, an even multiple of the tidal period. Such parallels at least hint at the common descent of all animals from a marine ancestor.

Other Tides

When oscillating tidal currents in the stratified ocean flow over uneven bottom topography, they generate internal waves with tidal frequencies. Such waves are called *internal tides.*

Shallow areas in otherwise open water can experience rotary tidal currents, flowing in directions that continually change and thus the flow direction (not the flow) completes a full rotation in 12½ hours (for example, the Nantucket Shoals).

In addition to oceanic tides, large lakes can experience small tides and even planets can experience *atmospheric tides* and *Earth tides.* These are continuum mechanical phenomena. The first two take place in fluids. The third affects the Earth's thin solid crust surrounding its semi-liquid interior (with various modifications).

Lake Tides

Large lakes such as Superior and Erie can experience tides of 1 to 4 cm, but these can be masked by meteorologically induced phenomena such as seiche. The tide in Lake Michigan is described as 0.5 to 1.5 inches (13 to 38 mm) or 1¾ inches.

Atmospheric Tides

Atmospheric tides are negligible at ground level and aviation altitudes, masked by weather's much more important effects. Atmospheric tides are both gravitational and thermal in origin and are the dominant dynamics from about 80 to 120 kilometres (50 to 75 mi), above which the molecular density becomes too low to support fluid behaviour.

Earth Tides

Earth tides or terrestrial tides affect the entire Earth's mass, which acts similarly to a liquid gyroscope with a very thin crust. The Earth's crust shifts (in/out, east/west, north/south) in response to lunar and solar gravitation, ocean tides, and atmospheric loading. While negligible for most human activities, terrestrial tides' semi-diurnal amplitude can reach about 55 centimetres (22 in) at the equator—15 centimetres (5.9 in) due to the sun—which is important in GPS calibration and VLBI measurements. Precise astronomical angular measurements require knowledge of the Earth's rotation rate and nutation, both of which are influenced by Earth tides. The semi-diurnal M_2 Earth tides are nearly in phase with the moon with a lag of about two hours.

Some particle physics experiments must adjust for terrestrial tides. For instance, at CERN and SLAC, the very large particle accelerators account for terrestrial tides. Among the relevant effects are circumference deformation for circular accelerators and particle beam energy. Since tidal forces generate currents in conducting fluids in the Earth's interior, they in turn affect the Earth's magnetic field. Earth tides have also been linked to the triggering of earthquakes.

Galactic Tides

Galactic tides are the tidal forces exerted by galaxies on stars within them and satellite galaxies orbiting them. The galactic tide's effects on the Solar System's Oort cloud are believed to cause 90 percent of long-period comets.

Misapplications

Tsunamis, the large waves that occur after earthquakes, are sometimes called *tidal waves*, but this name is given by their *resemblance* to the tide, rather than any actual link to the tide. Other phenomena unrelated to tides but using the word *tide* are rip tide, storm tide, hurricane tide, and black or red tides.

Bibliography

Archer, J.: *Coastal Management in the United States: A Selective Review and Summary*, Coastal Resources Center, Narragansett, RI, 1988.

Bauer, Steffan: *Handbook of globalization and the environment*, CRC Press, Delhi, 2007.

Berlin, G.L.: *Earthquakes and the Urban Environment,* CRC Press, Boca Raton, FL, 1980.

Boelaert-Suominen, S. and Cullinan, C.: *Legal and Institutional Aspects of Integrated Coastal Area Management in National Legislation*, FAO, 1994.

Bryant, D., E. Rodenburg: *Coastlines at Risk. An Index of Potential Development-Related Threats to Coastal Systems*, WRI, Washington, D.C., 1996.

Clark, J.: *Coastal Zone Management Handbook*, Lewis Publishers, New York, 1995.

Dolcemascolo, Glenn: *Environmental Degradation and Disaster Risk*, Asian Disaster Preparedness Centre, Sweden, 2004.

Drees, J.M.: *Coastal Dunes, Recreation and Planning. Proceedings of European Seminar*, EUCC, Leiden, The Netherlands, 1997.

Engelman, R., and P. LeRoy : *Sustaining Water Washington*, D.C.: Population Services International, New York, 1993.

Epstein, S. J. M.: *The Earthy Soil: Bombay Peasants and the Indian Nationalist Movement, 1919-1947*, Oxford University Press, Delhi, 1988.

Fox, Lisa L.: *Management Strategies for Disaster Preparedness*, Chicago, American Library Association, 1989.

Garcia Nova, F., Crawford, R.M.M., Diaz Barradas: *The Ecology and Conservation of European Dunes*, Universidad de Sevilla, Spain, 1997.

Ghost, K.C.: *Famines in Bengal, 1170-1943,* Indian Associated Publishing, Calcutta, 1944.

Hadfield, Peter: *Sixty Seconds that Will Change the World: The Coming Tokyo Earthquake*, Boston, C.E. Tutle, Co., 1992.

Hays, J.N.: *Epidemics and Pandemics: Their Impacts on Human History*, Santa Barbara, Calif.: ABC-CLIO, 2005.

Healy M.G. and Doody, J.P.: *Directions in European Coastal Management*, EUCC, Samara Publishing, Cardigan, 1995.

Jeftic, L., Milliman, J., Sestini, G.: *Climate Change and the Mediterranean—Environmental and Societal Impacts of Climatic Change and Sea-Level Rise in the Mediterranean Region*, Edward Arnold, London, UK, 1992.

Jones, P.S., Healy M.G. and Williams, A.T.: *Studies in European Coastal Management*, EUCC, Samara Publishing, Cardigan, 1996.

Kaur, Kanwaljit: *Handbook of Water and Wastewater Analysis*, Atlantic, Delhi, 2007.

Lewis, Peirce. *New Orleans: The Making of an Urban Landscape*. Cambridge, MA: Ballinger Publishing Co., 1976. Annotation

Merritt, John F. : *History at Risk, Loma Prieta: Seismic Safety & Historic Buildings*, California Preservation Foundation, 1990.

Nash, J.M.: *El Nino: Unlocking the Secrets of the Master Weather Maker*, Warner, Delhi, 2002.

Odum, Howard T.: *Environment, Power and Society*, New York, Wiley-Interscience, 1971.

Oliver, John E.: *Encyclopedia of world climatology*, Springer, Delhi, 2005.

Pernetta, J., and Elder, D.: *Cross-sectoral, Integrated Coastal Planning: Guidelines and Principles for Coastal Area Development*, Gland, Switzerland, IUCN, 1993.

Pilarcyzk, K.W.: *Coastal Protection*, Balkema, Rotterdam, The Netherlands, 1990.

Ross, David: *Ireland: History of a Nation*, Geddes & Grosset, New Lanark, 2002.

Salman, A., Berends H., Bonazountas M.: *Coastal Management and Habitat Conservation*, Marathon, Greece, April, 1993. EUCC, Leiden, 1995.

Trivedy, R.K. and Siddharth Kaul: *Low Cost Wastewater Treatment Technologies*, ABD Publishers, Delhi, 2001.

Vallega, A.: *Fundamentals of Integrated Coastal Management*, Kluwer Academic Publishers, UK, 1999.

Varley, Anne: *Disaster, Development Environments*, New York, J. Wiley, 1994.

Welp, M.: *Defining Integrated Coastal Zone Management for the Baltic Sea Region*, Springer-Verlag, Berlin, Heidelberg, 1999.

Wood, Charles A.; Jürgen Kienle: *Volcanoes of North America*, Cambridge University Press, UK, 1990.

Worster, Donald: *Dust Bowl: The Southern Plains in the 1930s*, New York, Oxford University Press, 1979.

Index

L

M

N

O

P

R

S

T

U

V

W

❑❑❑